The Whale Hunters

Other books by Robert Smith

Grampian Ways
Discovering Aberdeenshire
Valley of the Dee
The Granite City
The Royal Glens
One Foot in the Sea

The Whale Hunters

ROBERT SMITH

JOHN DONALD PUBLISHERS LTD
EDINBURGH

ISBN 0 85976 393 5

British Library Cataloguing in Publication Data
A catalogue record for this book is available from the British Library.

Typeset by IMH (Cartrif), Loanhead, Scotland
Printed and bound in Great Britain by J. W. Arrowsmith Ltd., Bristol

Preface

The story of the Scots who went to the Arctic last century to hunt the whale has never been put fully on record. Perhaps it never will be, for much of the material is buried in old logbooks, in faded diaries and long-forgotten letters, in whaling songs whose choruses were considered too crude to be published at the time, and even in scientific documents examining the killing of the whale.

This book is an attempt to set the record straight, at least in part: it is also a tribute to the Scots who went a-whaling, enduring unbelievable hardships in their search for the whale oil that would make their fortunes, *The Whale Hunters* comes at a time when the whaling era is drawing to an end, when only a handful of nations still pursue this Leviathan of the seas, From the whale-killing bloodbaths of a century and a half ago to cruise ships carrying tourists to the Arctic and Antarctic to catch a glimpse of Moby Dick . . . it is a long, incredible leap,

Among the many people who helped to make this book possible were a number of former whalers, men like old John Gray in Stromness, and Josie Manson, of Olna, Shetland, who served in South Georgia and now lives on the site of an old Norwegian whaling station at Olna. Another Shetlander who brought the whaling story up-to-date and provided some fascinating pictures was Stewart Gray, of Yell, who saw the last of the Antarctic whaling,

Austin Murray, of Wormit, son of a famous Arctic whaling master, gave invaluable assistance and also provided a number of photographs, Dr Graham Page, of Aberdeen, supplied dramatic pictures of the scene in South Georgia when the whalers had gone. Others who assisted photographically were Ian Tait, of the Shetland Museum, Bryce S. Wilson, secretary of Stromness Museum, John Edwards, curator of Aberdeen Maritime Museum, David Henderson, assistant keeper of Dundee Museum, and Dr

David Bertie, deputy curator of Arbuthnott Museum, Peterhead. Thanks also to Harry Smith and Mac in Canada, to my wife Sheila, and to many others who helped to put the pieces of this whaling story together.

August 1993 Robert Smith

Contents

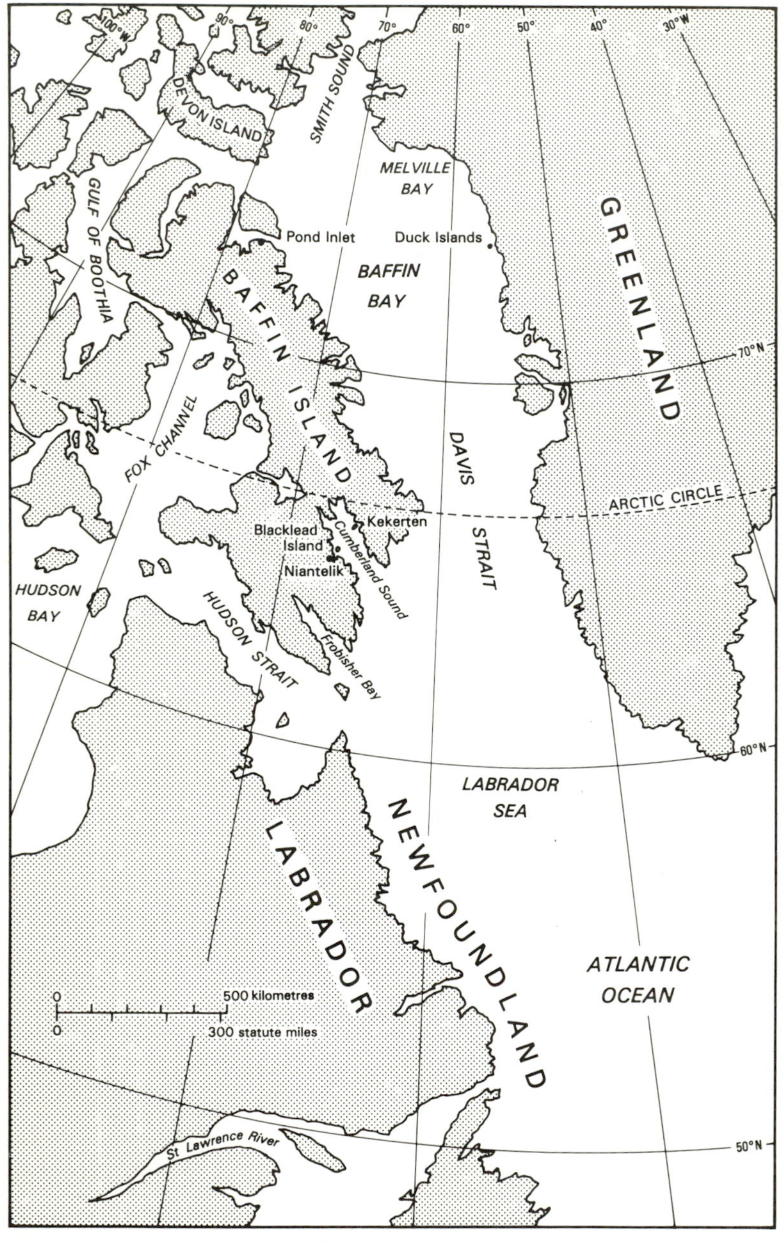

Location map

Introduction

> The discovery of new lands and new waters frequented by seals and whales has always been followed by their increased slaughter. In each case the unhappy animals are brought a stage nearer to extinction. Amongst the whales no species have suffered more than the Greenland whale—*Dr Robert W. Gray, of Peterhead, writing in April, 1920.*

In May, 1991, I flew from London's Gatwick to Denver, Colorado, a journey that took me north by Iceland to Angmadssalik on the east coast of Greenland, then south-west by the Davis Straits and Hudson Bay to Canada and America. We passed Godthab, on the west coast of Greenland, once familiar to whalers heading into the Arctic seas. 'This is Goodhab,' wrote James Cumming, surgeon of the Peterhead whaler *Eclipse* in 1830. 'All is drear and revolting.'

Thirty thousand feet below, pack ice lay over the Davis Straits like cracked icing on a wedding cake. I remembered Cumming's description of the area as 'a storehouse of storms'. This was the 'awful place' that whalers sang about in their shanties, a frozen world where men once hunted the 'whale fishes'. That month, the International Whaling Commission were meeting in Iceland to talk about the killing of whales, and to consider a proposal; to end a five-year moratorium on commercial whaling. To the alarm of conservationists, the Commission agreed to allow catches of Minke whales in the Antarctic, but postponed the fixing of quota until the IWC's meeting in Scotland in 1992. The decision was a blow to Iceland and Norway, who wanted to continue operating in their traditional whaling grounds.

Those whaling grounds lay below, along a ragged, icy coast whose inlets and bays carried the names of Arctic explorers like Frobisher, M'Clintock, Franklin, Amunsden, Ross and Parry. Then there was Penny's Land. Captain William Penny, the Aberdeen whale skipper, gave his name to it, as he did to Penny's Strait and Penny's Ice Cap, and if he had had his way there would have been more names from north-east Scotland on the Arctic map. In 1840, he had on board the *Bon-Accord* an Eskimo called Eenoolooapik, who had visited Scotland with him the previous year. While exploring the Baffin Bay area for new whaling grounds, they came upon an unmapped inlet which Penny named Hogarth's Sound, after William Hogarth, an Aberdeen man who had been kind to Eenoolooapik. Later, another whaling captain drew up his own map and re-named it Northumberland Inlet. To-day, it is known as Cumberland Sound.

One landmark was given the name of Cape Crombie, after the owner of Penny's ship, a harbour became Bon-Accord after the ship itself, and a place called Kingoua was renamed Davidson's Inlet in honour of another of Eenoo's Aberdeen friends. Many of the names were changed in later years. In the early 1700's, however, this part of the coast was virtually nameless. Whaling was confined to a narrow strip of open water along the west coast of Greenland, and the sea route to Baffin Island was blocked by a dense mass of ice known as the 'middle pack'. In 1817, the *Larkin* of Leith and an Aberdeen whaler, the *Elizabeth*, forced their way across the top of Baffin Bay and came out into what was known as the 'west water', where they found a tremendous run of 'fish'.

The Land of Desolation, the explorer John Davis called it, and it was here that the Greenland Bowhead whale once roamed the seas at will. Here, too, where towering icebergs poked long, needle fingers into the sky and clumps of rock reached up from the ice 'like the uplifted hands of drowning men', the Bowhead finally faced extinction. The year after the discovery of the 'west water' by the *Larkins* and *Elizabeth*, the Admiralty sent an expedition to the area to search for a 'north-west passage', using information supplied by the Scottish whalers. It confirmed the Scots' belief that whalers working their way south along the coast of Baffin Bay and the Davis Straits would find a rich harvest in the bays and inlets indenting the coastline.

The Bowhead had been pushed to the edge of these Arctic wastes by generations of whale-hunters, not least by the Basques, who had gone out from the Bay of Biscay into the north Atlantic as far back as the sixteenth century. 'We met a ship at sea', reported Davis, during his 1587 voyage, 'and as far as we could judge it was a Biscayan. We thought she went a-fishing for whales, for we saw very many.' For nearly a century after the discovery of the 'west water', the men who went a-fishing for whales engaged in an orgy of destruction that decimated the Bowhead population. This massive creature was also known as the Right whale. It was right for the hunters because it was slow, ponderous and easy to chase, and when it died its corpse floated on the water instead of sinking to the bottom of the sea, as other whales like the Rorquals did. It also had a heavy layer of fat, two feet thick in some instances, which could produce as much as 7,000 gallons of oil. It was, as one whaleman put it, 'a real oil-butt'.

To the Eskimo—the Bowhead was *Arveq*, the Big One. The Arctic waters were stained with the blood of this gentle giant. The whalemen plunged their lances into its huge body and called it 'tapping the claret bottle', and when a fountain of blood spouted from its blowhole they spoke about 'running up the red flag'. It was a 'flag' that was run up too often, leading to the virtual extermination of the Bowhead, but before this point was reached man's ingenuity had produced a murderous alternative. He had discovered how to hunt down and kill the species known as Rorquals, a name which came from the Norwegian *røyrkval*, meaning 'the whale with pleats'. The pleats were furrows along the animal's throat and belly.

The Rorquals included Humpbacks, Finbacks, Blues and Sei whales, whose speed had hitherto kept them out of reach of the whalers' harpoons. Now they were faced with the ultimate horror of the bomb harpoon. Herman Melville, in *Moby Dick*, wrote about the Right whale's 'high and mighty God-like dignity'. The whalers stripped it of both its dignity and its blubber, but the final agony was reserved for the Rorquals.

Ironically, it was meted out in the late nineteenth century by a Norwegian sea captain who claimed to have God's blessing on his work. Svend Foyn, the inventor of the bomb harpoon, declared that God had let the whale inhabit the waters for the benefit and blessing of mankind, and he considered it his vocation 'to promote these fisheries'. He did it with a harpoon that carried an explosive charge which blew up inside the whale's belly. It also contained four steel barbs which sprang open and anchored themselves to the whale's flesh.

Photographs of Foyn show him as a saintly old man. white-bearded and gentle, and, in fact, he was once described as 'a religious, good old man, respected and beloved by all who met him'. He wrote pious biblical quotations on each page of his log book, but his mind was often on more earthly things. In addition to inventing the bomb harpoon and building faster, steam-powered ships to hunt the whale, he also had an answer to the Rorqual's disturbing habit of plunging to the sea bed when it was killed. He invented the inflation lance, a hollow spear through which compressed air was pumped into the whale so that its corpse floated like a balloon on top of the water. It could then be towed to a shore station for processing.

Foyn's inventions revolutionised the whaling industry. Coming at a time when Arctic whaling was at its lowest ebb, it set in motion a fresh wave of killing that spread across the oceans of the world to the whaling grounds of the Antarctic. A new era was opened up for the whale hunters—and the day of the Bowhead had gone for ever. Many years later—in 1931—the last act in the drama took place when the killing of Bowhead whales was banned, although native hunters were excluded from the ban. By that time, there were said to be no more than a few hundred Bowhead whales in Baffin Bay, the Davis Straits and the northern reaches of Hudson Bay. In 1977, the International Whaling Commission took the final step of banning the killing of Bowheads by the Eskimos, but it later back-tracked on this and set them a small, tight annual. quota.

The Whale Hunters, which ranges from the Arctic to the Antarctic, spanning time as well as space, looks at whaling as it was and as it is now, and at what two centuries of slaughter have done to the whale population. Above all, it is the story of the Scots who hunted the whale. They came from ports all up and down the east coast, from the Northern Isles, and from Greenock in the west, where a handful of whalers operated. During different periods of the nineteenth century Aberdeen, Dundee and Peterhead dominated the whaling and sealing scene (Peterhead owed its success largely to its concentration on sealing), but smaller ports like Banff, Kirkcaldy and Leith also dipped their toes in the Arctic waters, some with little success. There were about twenty whaling ports in Scotland, half the total in the British Isles.

This is the story of men like the Grays of Peterhead, who left an indelible mark on the history of whaling; and of legendary whale masters like Captains John Parker and William Penny of Aberdeen, and Captain John Murray, of Dundee, who was the last of the Scottish whalers. It is also the story of the Orcadians and Shetlanders who went a-whaling, and who in many ways were the backbone of the Scottish whaling fleet. Not least, it is the story of the men from farms and fishertowns who left their homes believing that they would make their fortunes in the Greenland seas; men like 'Strong' John Hunter, the champion fighter of the whole Arctic fleet, or Eelie Bob, of the Peterhead whaler *Mazanthien*, who was immortalised in an old whaling song:

> Wauken up, Eelie Bob, or you're sure to be done
> In this year, eighteen hundred and fifty-one.

They fought starvation and disease, and they endured untold suffering, losing limbs and sometimes their lives. The history of whaling is laced with their agony. In *Whaling in Davis Straits*, written by an old sea master, Captain William Barron, there is a telling passage about the burial of a seaman at the Vrow Islands, near Upernavik. The frozen soil was only a few inches deep and his grave had to be dug with crowbars. All the ships had their flags at half-mast and the funeral procession consisted of thirty boats, each carrying six men. The boat with the coffin was towed to the land. After the service, a wooden headboard was erected. It carried the dead man's name, age, birthplace, and the name of his ship. Captain Barrow saw it a few years later and found it well preserved. 'I have seen other headboards fifty years after their erection in good condition,' he said.

Headboards like these can still be found in different parts of the Arctic. They are bleached and faded reminders of the cost in lives of that long-forgotten hunt for whale oil. Some never made it, but behind them came others, all nursing the hope that one day they would find their pot of gold in the Arctic seas:

> It'll be bricht both day and nicht when the Greenland lads come hame,
> Wi' a ship that's fu' o' oil, my lads, and money to our name.

CHAPTER ONE

Oil is King

It was a cold, gusty day when the whale ship *Christian* sailed from Aberdeen on Thursday, 10 February 1791. It left the port at noon, bouncing up the choppy waterway known as the Raik, past the pier built ten years earlier by the engineer John Smeaton. Smeaton's pier was known as the North Pier, and still is to-day, and it must have seemed to the crew of the little whaler that its long arm was signposting the way north . . . north to the Greenland seas and Jan Meyen, and farther north still to Spitzbergen, where the *Christian* hoped to pluck a fortune in whale oil from the sea. Instead, dogged by bad luck as well as by bad weather, she came home with an empty hold—a 'clean' ship. It was her last voyage to the Arctic. For a number of years there had been a scarcity of 'fish', as they called these awesome marine mammals, and in 1789, after catching only one whale, she cut short her voyage and returned home early. Her profitless trip two years later brought an end to her whaling days.

Today, huge oil supply ships tug at their moorings on Pocra Quay, the old Pow Creek burn, the fishermen's haven, where the *Christian* began her voyage. The only reminder of the city's whaling days is a rough granite tablet on the quay marking the site of a long-demolished blockhouse. Built in 1532, it served as a watch tower, a gunpowder store, a ferryman's home, and a boiling house for whale oil. Across the harbour, silos and oil storage tanks clutter the quays where a pattern of trawler masts once dominated the scene. Oil is king again, but it is a different kind of oil, and the ungainly oil vessels that chase the ghosts of the whalers up the navigation channel go about their business with scarcely anyone noticing their coming and going.

It was different in the days of Aberdeen's first 'oil boom'. When a whale ship left the port there were garlands, toasts, tears, and cheer, and people gathered on the North Pier to watch the whalers sail jauntily out of the harbour. The townsfolk were proud of their pier. It had become a promenade for Sunday strollers, a place where invalids came to breathe in the sea air, but as one writer angrily pointed out in a letter to the *Aberdeen Journal*, their nostrils were being assailed by an 'intolerable stench' from the boiling house on Pocra Quay.

It wasn't that Aberdonians were particularly sensitive; all up and down the coast, it seemed, people were holding their noses. Dundee found the smell of boiling blubber 'quite disgusting', while up in Shetland crofters complained bitterly about both the stench and the pollution of their beaches. Long after Peterhead had ceased to be a major whaling port there were grumbles about 'the horrible smell' from the local boiling house, although some Buchan folk actually believed that it was good for them. They thought it could cure tuberculosis, and TB sufferers regularly visited the boiler house to inhale the fumes. One writer euphemistically described the smell as 'a pretty strong perfume', and in Leith, where Peter and Christopher Wood owned one of four whaling companies, they spoke about the 'Woods' scent bottle'.

These east coast ports were the Blubbertowns of Scotland, but whatever stench emanated from them they were sweet-smelling compared to the huge settlement built in 1617 on Amsterdam Island in Spitzbergen, where the Scots were to hunt the whale more than a century and a half later. The Dutch sent 300 ships and 15,000 men to the Spitzbergen waters and the complex that housed them was called Smeerenburg—Blubbertown. As many as 1,500 Greenland whales were towed into its noxious harbour every season. The settlement lasted until the 1640's, then faded away.

The rancid odour of the Scottish boiling houses came later and lasted longer, lingering on for more than a century, but to those investing money in the risky business of whaling the genie locked in the 'scent bottle' was the wispy smell of success. Farley Mowat, author of a number of books on whaling, told of a latter-day American whale plant manager who dismissed complaints about the smell of boiling blubber with the comment, 'Who the hell gives a damn! That's the stink of money and it sure smells good to me'.

It smelled good to canny Aberdonians. The voyage of the *Christian* and its sister ships marked the coming of Scotland's first oil boom. Whale oil was essential to the economy. An advertisement in the *Aberdeen Journal* called for 'six tons of best boiled whale oil for the public lamps of this city'. Without it, the street lamps would have flickered and died. Wax candles lit the homes of the wealthy, but for the poorer classes there was a single tallow candle containing oil from whale blubber, with a rush or cotton wick. Oil was also needed for soap-making, dressing wool and vegetable fibres, and as an industrial lubricant.

Whalebone was a valuable ancillary product, much in demand by milliners and dressmakers. It was used to make parasols, hats, canes, fishing rods and suspenders—and corsets for the ladies. 'Urban conglomerations plugged into the corpse of the whale,' wrote Heathcote Williams in *Whale Nation*, and he was right, for it seemed as if half the nation's needs were met from the whale's carcase. Its uses increased as the years passed, and there was a good deal of truth in the joke that whaling plants put every inch of the whale to profit, except the spout.

Standing on Smeaton's pier, I thought of all the whale hunters that had followed the *Christian* up the Raik over the years . . . the 'bonnie ship', the *Diamond*; the doomed *Oscar*; the *Jane*, sailing into Aberdeen with one of the biggest catches ever landed at the port ('We'll go in to Jean Mackenzie's and buy a pint of gin, and drink it on the jetty when the *Jane* comes in'), the famous *Bon-Accord*, under Captain John Parker, the ill-fated *Middleton*, crushed in the ice; and the *Fox*, built in Aberdeen and bound for the Arctic to seek the missing Franklin expedition in the frozen backwaters of Cumberland Sound. There were many more. They pushed north to Orkney and Shetland, picking up extra crew at Stromness and Lerwick, and on to Greenland and Spitsbergen, reaching out into the Arctic.

It was in these northern waters in 1607 that the explorer Henry Hudson discovered what they called 'Hudson's Touches'—Jan Meyen Island—and, even more important, reported that he had seen a 'great store of whales'. It was like the crack of a starter's pistol. The English sent ships north to investigate, then the Dutch, and by 1612 the rush for oil—whale oil—was under way.

The Scots were slow off their mark in the whaling game. Aberdeen's first whaling company, formed in the mid-eighteenth century with two vessels, the *City of Aberdeen* and the *St Anne*, made little impression on an industry dominated by the English, the Danes and the Dutch. The company went out of business in 1775, but in 1783, boosted by a Government bounty, the Aberdeen Whaling Company came into being. The *Hercules* and the *Latona* were the first ships sent to Greenland and in 1787 the *Christian* joined them. This small triumvirate of whale ships became the advance guard of a Scottish whaling fleet that for much of the following century was to turn the Arctic into a vast 'fishery'.

Up in those northern waters, at latitudes of 77° N., the whalemen faced the danger of being trapped in the ice—'beset' was a familiar word in whaler's logs—and, day in, day out, had to contend with the gnawing, unrelenting cold. It ate into their very bones. George Kerr, the *Christian's* surgeon, recalled in later years that what remained strongest in his memory of the 1791 voyage was the sight of men carrying shovels laden with hot coals to thaw the spirit casks on the vessel.

Water and beer froze on the tables, drugs iced up in the medicine chest, and beef had to be cut into pieces with a saw. Kerr was unable to make entries in his diary because the ink had frozen over, and when he lay down on his bunk his pillow was like a block of ice. As the *Christian* went on beyond Jan Meyen hoar frost flew about in clouds, 'like summer fogs in

Scotland', dusting everything with a fine white powder. The rigging became stiff and almost unmanageable.

The Aberdeen ship saw only a handful of Scottish whalers as she pushed her way into the Arctic circle. She came upon the *Raith* and the *Neptune* from Leith, ill-fated ships, as it turned out, and she saw whalers from Dundee, where the death knell of the whaling industry was to be sounded a century later. She also spotted the little *Robert*, boisterously thrusting her way into whaling history. The *Robert* was the first Peterhead whaling vessel to be sent to Greenland, and by all accounts she should never have been there, for she was half the usual size of a normal whaling ship and scarcely big enough to carry the necessary number of boats.

The farther north they went the heavier the ice became. Kerr estimated one piece to be 'above a mile in length and about 40 feet thick'. Not many of the ships they met had had any success with the whaling, and Kerr noted in his diary that unless they had better luck before the end of the season it was likely that fewer than half of them would return the following year. He had little idea then that the *Christian* would not be going back.

There were vessels from London, Liverpool and Hull, but Danish and Dutch whalers were predominant. From 1675 to 1721 the Dutch had 5,886 ships at the whaling. Even by to-day's oil-boom standards their success was remarkable, for their catch totalled nearly 33,000 whales, worth a gross figure of £30,000,000. In early May, the *Christian* found herself among 71 whalers, most of them Dutch, magnificent vessels with three rows of cabin windows—'one storey for the harpooners upon the top, the middle for the master, and the surgeon and cook below'.

They had seen some of these vessels on their way north; so thick together, said Kerr, that they had 'the appearance of the distant view of a city'. Now, on the edge of the sea ice, they lay together side by side, held by their ice anchors, showing 'as little motion as in the harbour of Aberdeen'. The ice stretched as far as the eye could see, ten feet high above the water, broken into pieces about a quarter of a mile to half a mile in diameter. Gradually it closed in on them, and by the evening of 21 May they were completely beset. Some of the whalers were able to make their way out of the drifting ice, but from the deck of the *Christian* 38 vessels could still be seen locked in it.

On Thursday, 22 May, word came that the *Neptune* had been sunk. That same morning, the *Christian's* crew watched a line of men crossing the ice about three miles away, carrying a red ensign with a white cross on it. They thought they were Danish, but when the *Christian* sent provisions over to them the 'Danes' turned out to be members of the *Raith's* crew. They had spotted a dead whale on the ice and had gone to collect it, but when they were clear of their ship the ice broke off from the main mass. It drifted away, taking the *Raith* with it. The Leith whaler survived, only to be wrecked some twenty years later.

The whalers lived in fear of having their ships 'squeezed' by the ice. They lay in their bunks and listened to the vessels cracking and groaning under the constant pressure, and they were kept awake at night by intermittent explosions, like cannons being fired. The noise was caused by huge pieces of iced being thrown into the air. If one of them had landed on a ship it would have had the effect of a stone falling on an egg-shell. The floes pressed closer around the *Christian*, and the men hauled out their bedding and made ready to abandon ship, but the pressure eased off. 'What a fine school for an impatient man!' wrote George Kerr. 'Unable to stir a foot from the place, we drifted about at the pleasure of the wind.' There were reports that ten vessels had been sunk, but nobody knew for sure. They had given up any hope of fishing; all they wanted was to get out of their ice-bound prison.

Despite their fears, there were times when life became more tolerable, when everything seemed to be softer and gentler, the ice and snow less savage, the wind less bitter; when a kind of tranquillity descended on the scene. It was as if Nature had relented a little, drawing back from its terrible onslaught.

Paths criss-crossed the ice where men had gone from one ship to the other; 'gamming', they called it, exchanging gossip, talking about home, dulling each other's anxiety about their future. The tracks reminded Kerr of the roads going from one house to another at home, particularly at Christmas. He heard the ringing of bells on the Dutch whalers, calling the crews to prayer, and when he closed his eyes it was like hearing the familiar tolling of church bells on a Sunday morning.

Religion comforted some men, drink blurred the fears of many. 'The sailors walk in dozens over the ice to see one another and are scarce ever free of liquor,' wrote Kerr. The surgeon also went visiting, but more soberly. He crossed the ice to the Dutch ship *Pro Partia* and was welcomed on board by a bluff Falstaffian figure who turned out to be the Captain. He was, said Kerr, 'a great overgrown fellow, scarce fit to totter along the deck'. Fifty years old, he was a veteran Arctic whaler, having sailed to Greenland every year since he was ten years old.

After coffee, Kerr and the Captain's son played the fiddle together. The Dutchman liked Scots music, and the sound of reels and schottisches echoed across the ice. Kerr found himself caught up in more musical entertainment when he left the *Pro Partia*. On his way back to the *Christian* after treating the gangrenous feet of an English sailor he passed the *D'Lillie* of Hamburg and saw five Danish surgeons standing on the deck singing out like a barber's quintet. Kerr said they sung English airs better than he had ever heard them on the stage.

But these were brief interludes in an otherwise desolate existence. Kerr thought they would never get out of the ice. When two Dutch vessels were wrecked, men from the *Christian*, trying to salvage wood and ropes from them, were attacked by the Dutchmen, and when the *Christian* was holed by ice they had to fight off Dutchmen who had come to plunder their vessel.

On 29 June, Kerr awoke to find that the ice had drifted two miles away from the ship. It soon closed in again. It was a stop-go kind of progress, the ice slackening and letting them move towards open water the wind changing and halting them. 'I scarce know whether to give up hopes of getting out or not,' he wrote. On 21 July, however, they were able to raise sail, and by late afternoon the *Christian* was out in open water.

The *Christian* reached the coast of Norway on 8 August and on the following day set course for Buchan Ness. They saw Scotland on 11 August and at seven o'clock next morning the familiar kirk steeple at Peterhead came into view, telling them that their journey was almost over. By noon they were within sight of Aberdeen, but the sea was too strong for any boat to come out and they anchored off the mouth of the Don. On 13 August they entered Aberdeen harbour—home at last, a 'clean' ship, with nothing to show for their voyage.

Across the channel from Porca Quay, where the *Christian* berthed, lies modern Torry, where there is a street called Oscar Road. Not many people know how it got its name, and to find its origin you have to go back to 1 April—April Fool's Day—in 1813, when five whalers lay at anchor in Greyhope Bay. Two were the *Hercules* and the *Latona*, which were still in the hunt for whale oil, and the others were the *St Andrew*, the *Middleton*—and the *Oscar*.

The weather was unsettled with a rising wind and thick snow flurries scudding across the harbour. The *St Andrew* and the *Oscar* weighed anchor and attempted to clear Girdleness and run to the south, but heavy seas and gale-force winds forced them back. The *Oscar*, after dragging her anchor, was driven ashore behind the breakwater, at the south side of the harbour. A huge crowd on the shore watched in horror as the vessel was pounded to pieces. Only two men survived.

The year of the *Oscar* disaster started out as a year of promise, with fourteen Aberdeen ships sailing to Greenland waters and the Davis Straits—'as fine ships as ever put to sea', declared the *Aberdeen Journal*. Yet of the five ships lying in Greyhope Bay on that April Fool's Day, only one escaped destruction either then or later. The *Oscar* went to its doom even before it had sailed for the Arctic, and in the same year the *Latona* was struck by ice in the Davis Straits and went down in fifteen minutes. The *Middleton* was wrecked in 1830, and the *St Andrew* was lost with all hands in 1861. The only survivor was the *Hercules*, which ended its days in the Atlantic timber trade.

Whaling was a dangerous and chancy business. There were good years and there were bad. Some ships had luck on their side, others had not. Even the big names in whaling struck poor patches, sometimes coming home 'clean'. The *Bon-Accord*, under the legendary Captain John Parker, had a bad run in 1836, a year in which the crew never saw a whale, didn't even get a seal, and had to beg enough oil from another vessel to give themselves light until their ship reached home.

There were also the whalers that didn't come back. Twelve Aberdeen whalers were lost between 1813 and 1840 and out of a total of twenty ships whaling from the port between 1800 and 1840 no less than 60 per cent were lost. Four were lost in 1830, a disastrous year in which 19 of the 91 British ships at the Davis Straits were lost in the ice and 21 returned 'clean' without fish. It was the year that marked the start of Aberdeen's decline as a whaling port, a downward trend that was hastened by the coming of gas.

While Aberdeen slipped from its position in the whaling league table, Peterhead and Dundee whale ship were still returning to their home ports with their holds full of blubber and bone. In 1791, the *Christian* pointed the way to what promised to be a golden era for the whale hunters, and to some it was. To others, it brought financial disaster, and its toll in human misery and suffering was immense. More than that, it turned the Arctic seas into a slaughterhouse in which the Greenland Bowhead whale was hunted to the point of extinction.

It was well into the twentieth century before the last whaler sailed to the Davis Straits from a Scottish port, but Aberdeen, which had produced some of the finest whaling masters in the country, disappeared from the scene long before that. There were no more whale ships in the port after 1858. All that remained were the legends.

CHAPTER TWO

A Silver Penny

A pair of silver snuff boxes can be seen on display at Aberdeen's Maritime Museum. One was presented to Captain John Parker by the Bon-Accord Whaling Company in recognition of his long service as master of the whaler *Bon-Accord*, the other was a gift to Captain William Penny from the crew of the *Lady Franklin.* They were the giants of the whaling industry in Aberdeen, their whaling activities straddling almost the whole of the nineteenth century.

Parker and Penny, like all Arctic shipmasters, had to fill a variety of roles. Their skill and judgement often meant the difference between life and death. They had to be natural leaders of men as well as skilled seamen, geologists as well as naturalists, geographers as well as explorers. Many of the bays and inlets on the shore of Baffin Bay were discovered and named by whalers.

They also had to be highly skilled in the art of ice seamanship, which was a vastly different thing from ordinary seamanship. Towing and tracking were the normal ways of getting a ship through the ice, but sometimes the captain had to fall back on older methods. 'Overing' was one of them. It was done by creating a human see-saw, with the crew running *en masse* from one side of the ship to the other to bring about a rolling movement which broke up the ice. 'Milldolling' was an even older method. A boat was slung under the bowsprit by a tackle and rolled from side to side so that the ice was cleared ahead of the ship. If this didn't work, the boat was lifted up and dropped on to the ice to smash it up. If *that* didn't work, a boy was put in the boat to give it extra weight!

Both Perry and Parker were hard taskmasters. Penny could be dour and unswerving, but he held the respect of his men, as was seen when the crew of the *Lady Franklin* presented him with a snuff box 'as a token of regard'. Parker was a tough disciplinarian, even to the point of ruthlessness. On one occasion, when the *Bon-Accord* was anchored in the ice, the ship's sailmaker went off to visit a friend on another vessel some distance away, but didn't bother to ask permission.

When a breeze blew up, clearing the ice, the *Bon-Accord* got under way. The missing man was seen crossing the ice in an attempt to catch up with the ship, but the gap became bigger and the sailmaker grew smaller and smaller. Finally, he was seen only as a speck on the horizon. A deputation from the crew went to Captain Parker and pleaded with him not to leave the man to his certain death, but he refused to listen. Luck, however, was on the sailmaker's side. The *Bon-Accord* became entangled in the ice and he was able to catch up with her after a sixteen-hour chase.

Drink was often a problem, but Parker had his own way of dealing with it. When the *Bon-Accord* put into Stromness on the way home from a whaling trip, he gave instructions that nobody should leave the ship. Later, he discovered to his fury that every man was ashore. They boarded the ship again in the morning, most of them drunk. Captain Parker was waiting by the gangway with an iron pin in his hand, and when anyone passed him with a lump under his jacket he tapped it with the pin. Liquor ran in a stream down to their feet.

The last to return was the seaman who had incited the 'mutiny'. Parker refused to let him on board the ship, despite the pleas of the Orkney watermen who had taken him there. There were no regular sailings in those days and the stranded seaman had to wait weeks before he got passage home. Parker, however, arranged for the man's family to be paid his wages while he was stuck in Orkney.

Parker could be as stubborn with his bosses as he was with his men. Whales or no whales, he usually set off for home by 1 October, but in 1835, after a poor year's fishing, a meeting of owners decided that no vessel with less than 100 tons of oil should leave the whaling grounds until 15 October. Parker immediately resigned, making it clear that decisions like that would not be made by owners sitting comfortably in their parlours at home—*he* would decide when his ship left the ice. The owners had no desire to lose someone of Parker's calibre and he got his way.

Like all whalers, however, there *were* times when the *Bon-Accord* was beset in the ice, often for weeks, and Parker had a novel way of keeping his crew fit—he made them play 'Follow My Leader'. Up the rigging and back stays and out to the yard arms they went—'wherever it was possible for a monkey to go we had to follow,' said one crew member. When they weren't scrambling up among the topsails they were holding races, playing leapfrog or kicking a football.

The Duck Islands were well-known to Arctic whalermen. To the men of *Bon-Accord* they had a special significance. When they were blocked in by ice near the islands the watch on deck had to go ashore every morning to rob the ducks' nests for breakfast eggs. The nests were holes scraped in the sand where there was a patch clear of snow, and the ducks fought so hard to defend their eggs that the men had to carry sticks to protect themselves. The ducks got vigorous support from colonies of Burgomasters (Glacuous gulls), whose wings could badly hurt a man. These birds were so fond of blubber that when a whale was being flensed they descended on it in their hundreds. It took two or three men to fight them off.

Occasionally, the whalermen hunted and shot a white bear, and the boys on the ship were given the job of tanning the skin. This was done by putting it in a canvas bag full of sawdust and threshing it with a flail. The sawdust, which absorbed the oil, was frequently changed, and when it became as dry as a bone and as soft as a glove it was presented to the captain.

John Parker commanded the *Bon-Accord* from 1813 until 1840, when she was sold to Hull owners. He carved a niche for himself in Scottish whaling history, but his name—or, at any rate, the name of his ship—might have found a place in Eskimo lore if things had been different. Lewis Middleton, an Aberdeen man who sailed with Parker and later commanded his own ship, visited an Eskimo settlement about three miles from where the *Bon-Accord* was trapped in the ice. On his way back he met an Eskimo woman—'an old crone', was how he described her—who had given birth to a baby two miles from any shelter. She had wrapped him in her husband's sealskin jacket, and she planned to call him 'Bon-Accord'. Sadly, when Middleton returned the following year he was told that Baby Bon-Accord had died.

The Eskimos from the settlement were invited to visit the whalers. They travelled twenty or thirty miles to the ship and when they reached the water's edge they stood and shouted, 'Bon-Accord! Parker! Bon-Accord! Parker!' They were taken on board, where they ate ravenously from the ship's food stores, danced to a musical instrument called the Humstrung (something like a guitar), ate again, and at midnight announced that they were now going home.

In his *Whaling Recollections*, Middleton mentioned a revolting whaleskin-blubber sandwich which the Eskimos ate with great relish. 'They are very fond of blubber,' he wrote, 'and the greatest treat we could give them was to allow them to get their fill of whaleskin and blubber, which they ate just as we do bread and butter.' The civilised Scots had *their* own favourite Arctic dish—'a very eatable sea pie' made from water fowl. They were sometimes given a special treat. Their normal food consisted of hard biscuits, beef, pork, barley or pea soup, potatoes and batter pudding for dessert twice a week. The treat came on the birthdays of Captain Parker's two daughters, when the Captain celebrated by issuing a special helping of batter pudding with raisin in it, and a glass of grog.

Lewis Middleton got an even tastier titbit—a pound of candy made by a 'sweetie wife' in Aberdeen called Jenny Milne. The candy, sent out be a relative, was carried all the way to the Arctic on the *Hecla*, an old discovery ship, by an up-and-coming young mariner whose name was to become known wherever whalers met—William Penny. It was in those early days that Penny established a firm friendship with the *Bon-Accord's* master, John Parker, but he could never have anticipated that one day Parker, a shareholder in the company that owned the *Bon-Accord*, would offer him command of Aberdeen's most famous whaler to carry out exploration work.

During Penny's trip on the *Hecla*, ice began to close in on the whaler, and Penny made his men tow the vessel past the rest of the fleet—and then headed north. This piece of seamanship was regarded as an act of folly by veteran whalers. Weeks later, when the ice opened up and freed the trapped ships, it was too late for them to follow the *Hecla*. They turned for home with empty holds, but the *Hecla* came back with a full cargo—a small fortune for her owners. From then on, Penny was accepted as a skilled and competent whaleship master.

He came from that cradle of legendary whalers, Peterhead. Born in 1809, he was the son of a whaling master, William Penny, Sen., who for a time commanded the first *Eclipse*. Young Penny went to sea with his father as an apprentice on the *Alert* in 1821, when they brought back eighteen whales. In 1832 he became mate on the *Traveller* of Peterhead under the command of Captain George Simpson, one of Peterhead's top whaling skippers.

Whales were scarce in the 1832 season and it was on Penny's suggestion that Captain Simpson took his ship up Lancaster Sound. The result was that when the *Traveller* returned to Peterhead it was the third most successful British whaler, having caught 38 whales. Even in the early years of his career, Penny was constantly thinking of how to open up new fishing grounds. The ailing industry badly needed a shot in the arm and Captain Simpson, along with Penny, decided to look at the possibility of hunting whales in uncharted inlets in the south. Penny, with two Eskimo navigators and a whaleboat crew, was sent to investigate a bay where Eskimos had reported an abundance of whales. Unfortunately, high winds forced them to turn back.

But exploration was in Penny's blood. In 1835, at the age of twenty-six, he was given command of his first ship, the 220-ton barque *Neptune*. He began to think again about new fisheries, including a large bay south of Cape Walsingham known to the Eskimos as Tenudiakbeek. His initial inquiries met with little success, but in September, 1839, he met an Eskimo who had been born on the shores of Tenudiakbeek Gulf in a village called Keimooksook. His name was Eenoolooapik. Penny decided to take him back to Scotland so that he could win support from the *Neptune's* owner, William Hogarth, and from the Government, for an expedition to the region.

The arrival of Eenoo, or Bobbie, as he was called, caused a sensation in Aberdeen, but the Eskimo, although popular with the canny Aberdonians, was not impressed with the climate—'Too much cough', he said, after falling sick and nearly dying. Meanwhile, Penny had difficulty in getting his plans under way. Aberdeen, which once boasted fourteen whaleships, had only three in 1839 and Hogarth decided to take the *Neptune* and *St Andrew* out of the industry. It looked as if Penny's ambitious plans had fallen apart, but in 1840 he got his ship—the *Bon-Accord*.

The whaler sailed on 1 April 1840, passing Cape Farewell by early May. In July, they were north of Cumberland Strait, where they found an inlet which Eenoolooapik confirmed was Tenudiakbeek. Penny named it Hogarth Sound after the *Bon-Accord's* owner. There was no sign of whales, but the Eskimos said they would appear when winter came.

The *Bon-Accord* sailed south to Eenoo's birthplace at Keimooksook, where he left the ship. Alexander McDonald, a member of an Edinburgh natural history society, who was on board the *Bon-Accord*, said that Eenoo showed no emotion when they parted; there was no 'longing, lingering look behind'. Penny's friendship with Eeoolok didn't end there, however, for he visited his old friend on every voyage until the Eskimo's death.

After their leave-taking at Keimooksook, Penny returned to Hogarth Sound (it was later re-named Cumberland Sound) and found a large number of whales entering the inlet, just as the Eskimos had forecast. The *Bon-Accord's* men harpooned two, but lost them, so that in the end they left the Davis Straits without having caught a single whale. The *Bon-Accord*, returning to Aberdeen 'clean' was sold to Hull, the port was left without a single whaler, and Penny was made redundant.

Back home, Penny married an Aberdeen girl, Margaret Irvine, and they set up home in Marischal Street, where such distinguished citizens as the writer William Kennedy and the artist William Dyce had their residences. In 1844, Penny was off to the Arctic again on the *St Andrew*. But another challenge lay ahead. In 1843, Sir John Franklin set off with two ship, the *Erebus* and the *Terror* to search for a 'north west passage'. Nothing was heard of the expedition and whalers were asked to keep a look out for it. Penny did what he could to locate the missing ships, but his efforts upset the owners of the *St Andrew*, who thought he was spending too much time on it. Penny gave them a short reply—he resigned.

His new vessel was a Dundee whaler, the *Advice*. He made only one voyage on her, and on this trip he again tried to contact the Franklin ships. When he returned to Aberdeen, he wrote to Lady Franklin and the Admiralty and offered to lead a Franklin search expedition. The Admiralty were impressed by this 'daring but prudent' whaling master, and Lady Franklin liked the man she called her 'Silver Penny'. He was given two ships—HMS *Lady Franklin*, built in Aberdeen by Walter Hood, and HMS *Sophia*. Penny, however, fell out with Captain H. T. Austin, who commanded four other ships taking part in the search, and his request for command of another expedition in 1852 was turned down.

The decision angered Penny. He had always nursed the thought that he was lower in the social scale than the gold-braided Navymen. When he wrote about his treatment by the Admiralty he described himself sarcastically as 'a speculative adventurer', a 'NW trapper' (he spelt it traper), and a whaling master who had been robbed 'of the well-earned applause of the country'. He said that influence had 'placed him on board an Expedition and called him Captain', and influence had 'assembled Bishops and Dukes and the intellects of the Land to hear a paper read of a pleasure voyage round Baffin's Bay attesting the truth of Baffin & Bylots early discoveries.' His reward, he declared bitterly, was that his children had been 'robbed of their Patrimony by his being thrown out of employment'.

Penny was out of the Franklin search, but, oddly enough, it was an Aberdeen-built ship that played a major part in the closing chapters of the Franklin story. The Government pulled out of their commitment to the search and Lady Franklin fitted out an expedition at her own cost. The yacht *Fox*, adapted for Arctic conditions by the shipbuilders, Alexander Hall, was put under the command of Captain Francis McLintock. She sailed on 1 July 1857. Lady Franklin travelled north to see her off, and hundreds of people turned out to cheer her on her way. In 1859, McLintock's men discovered the log of the last days of the *Erebus* and the *Terror* and the Franklin drama was finally over.

Meanwhile, the *Lady Franklin* was finding a new role. In 1853, the Aberdeen Arctic Company was formed, with William Penny as general superintendent. The objectives of the company were 'the application of the power of the auxiliary screw to whaling vessels, the establishment of fishing settlements in the bays and inlets of Davis Straits, and especially of founding a fishing and mining colony in an inlet known as Northumberland Inlet, or Hogarth Sound, discovered by Captain Penny'. Later, the introduction of steam whalers was postponed and it was decided that the company's ships would winter in the sound. Wintering out had always been one of Penny's great dreams. When the Government granted the new company a piece of land on the shores of Hogarth Sound it suddenly became possible. The whalers that were to do it were the *Lady Franklin* and the *Sophia*.

They sailed from Aberdeen on 9 August 1853, with Captain Penny in command of his old ship and the *Sophia* under Captain George Brown. The crew of both whalers spent the winter on board the *Lady Franklin* and they returned to

Aberdeen in the autumn of 1854 with 260 tons of whale fins. With this new pattern of whaling in the Davis Straits firmly established, prospects for the Aberdeen Arctic Company were bright, despite the fact that the British whaling industry was on the decline. Penny arranged for houses to be built at Kekerten, an island north of the sound, and at Nuvuk, on the northern shore.

Throughout his years in whaling, Penny had always been conscious of how their activities could affect the lives of the Eskimos. The growing numbers of wintering whalemen, both British and American, put their very survival in jeopardy. In 1857 he took his wife and young son with him to Cumberland Sound. The *Aberdeen Herald* reported that Mrs Penny 'not only successfully braved the rigours of the Arctic, but by intercourse with the natives was of great service to the expedition.' Also with them was Brother Mathias Warmow, a missionary from the Moravian Church. Warmow was disturbed by the conditions in which the Eskimos lived and alarmed at seeing them dressed in European clothes and imitating Europeans. 'They were undoubtedly better off in their original state.' he wrote.

Warmow thought there was a danger that the Eskimos might 'speedily die', and, in fact, there had been a sharp drop in the Eskimo population. In 1840, when Penny discovered the Sound, the figure had been 1,000, but in 1857 it had gone down to 350. Penny made strenuous efforts to have a permanent missionary settlement established there, but after receiving Brother Warmow's report the Moravian Church decided against it. It marked the end of the Aberdeen captain's involvement in the development of Cumberland Sound.

Yet Captain Penny's eyes were still fixed on new horizons. Ironically, his final project was to put Aberdeen's rival, Dundee, on the road to becoming Britain's leading whaling port. His own company had lost interest in the introduction of steam whalers, so Penny went to the Dundee shipbuilding firm of Alexander Stephen and Sons and offered his services. The result was that he was engaged to superintend the building of a wooden barque-rigged steamer, the *Narwhal*, which was completed in 1859. In the next twenty years the *Narwhal* did well at the whaling, which may have had something to do with the fact that the whaler had a piper on board to spur the crew on to greater efforts. The *Nova Zembla* also had a piper in her crew.

Penny commanded one of the new Dundee steamers, the *Polynia*, on her maiden voyage in 1861. Not surprisingly, his work on the Dundee ships upset the directors of the Aberdeen Arctic Company and in 1862 he was fired. His whaling career ended where it began—in Peterhead. In 1863 he took command of the Peterhead whaler *Queen* and sailed to Hudson Bay. He wintered in Hogarth Sound, the scene of his greatest success, and in the spring of the following year he left the Arctic for the last time. He died on 1 February 1892, at the age of eighty-two.

CHAPTER THREE

Seal Slaughter

The boats they pulled to leeward,
Went skipping o'er the sea,
And killed this noble whale-fish—
Another Jubilee!

The golden Jubilee of Queen Victoria was celebrated in the Greenland seas on 21 June 1887, by the killing of a 57 foot long female whale. It was the largest whale ever caught by Captain David Gray, of the Peterhead whaler *Eclipse*, and a Shetland seaman on the *Hope*, captained by David Gray's brother, John, wrote a special Jubilee song to mark the 'glorious day'. He declared that when they got back to Lerwick they would march through Commercial Street and 'sing the Jubilee'. The whale yielded 27 tons of oil. Its jawbone, which measured close on 20 feet, was sent to South Kensington Museum in London.

Other whalers with the *Eclipse* were less lucky. The *Erik*, a Dundee ship, caught only a small whale—it 'measured three feet three', said the Jubilee song. 'Three-feet-three' was the length of the longest strands of whalebone out of the 365 pieces in the whale's head, which meant that the 'fish' was at least 20 feet long. The *Hope* got nothing.

The Shetland poet grumbled about his own ship's lack of success—'the *Hope* has none, and none shall get'. As far as he was concerned it had been a 'dreary voyage', and he felt that there was a lack of fair play on the *Hope*. He vowed that in future:

We'll never ship for one-and-three,
Because we didn't get fair play
The year of Jubilee.

'One-and-three' was the sailors' oil money. They boasted that when they were paid off on shore they would have 'plenty of brass and a bonny lass', but 1s. 3d. a ton was scarcely the sort of 'brass' they had expected when they signed on and went off to the Arctic to make their fortunes.

The song, entitled 'The Eclipse', was written four years before Captain David Gray retired. Among the late nineteenth century whaling captains he was considered the most skilful. He was also widely-known for his scientific knowledge, for he regularly contributed papers to scientific journals on Arctic natural history and the habits of whales. Yet it took almost a century for his home town, Peterhead, to honour him. In 1991, the Blue Toon named a street after him—Captain Gray Place.

For many years, he lived near the harbour in an imposing house called 'The Links'. Perhaps the ghost of the old sea-dog still peers out from the windows of 'The Links', looking across the water to the clutter of oil ships in the Harbour of Refuge. The days of whale oil have long since gone, but the Blue Toon has been caught up in another kind of oil boom. Beyond the curiously-named Queenie, a channel separating Keith Inch from the harbour, the buildings are given over to oil-related activity. There were houses there at one time, and folk who lived in them were known as Queenie Arabs. David Gray was a Queenie Arab. He was born in Castle Street, Keith Inch, on 29 October 1829. To-day, not far from his birthplace, there is a link with the old whaling days—a plaque marking the construction of a modern quay built to cope with the sea-going traffic. Its name—Blubber Box Quay.

The Grays and their relatives were involved in Arctic whaling longer than any other family in the British Isles. This north-east whaling dynasty had its roots in the late eighteenth century, when Captain A. Geary, who commanded the port's first whaleship, the 169-ton brig *Robert*, married Barbara Gray, daughter of Captain David Gray, Sen. Gray took over the *Perseverance* in 1811, and in 1826 his son, John, became captain of the *Active*. John had three sons—John, David and Alexander—and while all were successful whalers it was David who left an indelible mark on the whaling scene.

In 1844, when he was only fourteen, he went to sea with his father on board the *Eclipse*, a ship that had been taken over from one of John Gray's relatives, Captain John Suttar, who was known as 'Goodie' or 'Psalm-Singing' Suttar. Some twenty years later, David Gray took command of another *Eclipse*, a steam whaler that was to write its name across the pages of Greenland whaling history. He was following a tradition set by his father on the *Old Eclipse*, which was invariably top ship in the Peterhead fleet. In 1854, it gave its name to an inlet in the Davis Straits—Eclipse Sound.

Although the reputations of both Captain David Gray, Sen., and his son, John, stood high in the whaling world, David Gray was well able to match them. He had his own style, which was a curious mixture of boldness and native Buchan caution. It is significant that in half a century at sea he never lost a ship and never ran aground; his skill as a seaman and navigator was unquestioned. Unlike his father, he never went to the Davis Straits, working only on the Greenland grounds.

During his forty-three years as a whaling captain, David Gray killed 198 Bowhead whales and 168,956 seals. To-day, conservationists would turn a jaundiced eye on such slaughter, and Gray's son, Dr Robert W. Gray, who sailed with him on eight voyages, came to regret the wholesale killing of whales and seals. Now, when the International Whaling Commission talks about steps to protect the world's whale population, it is worth remembering that David Gray himself suggested a moratorium on whaling—for reasons that would bring a hollow laugh from conservation bodies.

He felt that the killing of Bowheads had nothing to do with the apparent decline in the whale population. Steam power was at fault. Gray thought that whales could detect the sound of an engine a long way off and, given this warning, went out of their way to avoid the whalers. In other words, it wasn't that there were fewer whales, it was simply that they had become more elusive. He believed that if there was a moratorium on whaling for a number of years, the whales would again become less wary. Although his proposal was never taken up, there was at least a small grain of truth in his argument. Many of the steam whalers were still fully rigged, which allowed them to operate entirely under sail, with their engines silenced, when on the actual fishing grounds.

Blame for the extermination of the Greenland whale can be laid at many doors, but the Gray family were far from guiltless. In 1838, after British whaling fleets had suffered three disastrous seasons, the Peterhead fleet of ten ships took eighty whales and 28,708 seals. The top ship was the *Old Eclipse* under Captain John Gray, with 22 whales and 5,500 seals. Most of the whales were nursery whales, not long weaned. They were no more than 30 feet long—the kind of 'three-feet-three' caught by the *Erik* in Jubilee year—and they were easily killed. The nursery whales were generally found north of 80 degrees. In the late eighteenth century, when ships were being strengthened to penetrate the ice, nursery whales fell prey to the whalers. This killing of young fish was the first step towards wiping out the Bowhead population.

In tracing the decline of the Greenland whale, Dr Robert Gray singled out 1814 as one of the most disastrous years. 'Hundreds of the adolescent animals were killed and much harm done to the species and to the fishery,' he said. Interestingly, it was a member of the Gray 'clan' who helped to make it a disastrous year. Captain John Suttar, in command of the Peterhead whaler *Resolution*, returned home with the largest catch ever taken by a Scottish whaler—44 whales, giving 299 tons of oil. Only nine were sizeable fish, the rest were small.

Basil Lubbock, in his book, *The Arctic Whalers*, said that in 1814 there was 'a greater slaughter than ever of young fish in the Greenland seas.' Even the Grays were shocked by the wave of killing, and particularly by the part played in it by their 'Psalm-Singing' relative, whose cargo realised £9,568 for oil and £1,100 for bone. It is said that the slaughter of baby whales by 'Goodie' Suttar put him out of favour with the family.

In later years, having laid aside his theory about elusive whalers and noisy engines, Captain David Gray wrote a report to the Scottish Fishery Board which said:

> The scarcity of whales now is not so much owing to the numbers killed in Greenland and Davis Straits, although there has no doubt been a vast number of them killed from first to last. It is more owing to the way the earlier whale fishers conducted their business in killing off the young whales before they were old enough to reproduce. In this way a large number of the old ones died out and no young ones were left to grow up and take their place.

The belief that small was beautiful—and profitable—was perpetuated in the 1830's by the hunting of 'Chaney John'. Chaney John was the whalers' name for the Bottlenose whale, which until that time had been largely ignored because of its size. The Scottish whalers also called it the 'Botley'. In 1877, the *Jan Meyen* of Peterhead killed ten 'Botleys' and in 1880 Captain David Gray killed thirty-two. By then the Bowhead had virtually disappeared and Gray was the first whaling skipper take up Bottlenose fishing as a serious commercial proposition. In 1833 he killed 200 Chaney Johns and said that if his men had not been new to the work they would have killed many more.

These small whales, as well as being 'skinners', were fast-moving and difficult to kill with the hand harpoon, but the bomb harpoon changed all that. David Gray described them as 'very unsuspicious'. They came close alongside the ship, poking inquisitively around it until their curiosity was satisfied. 'The herd never leaves a wounded companion so long as it is alive, but they desert it immediately when dead, and if another can be harpooned before the previous struck one is killed, we often capture the whole herd. They come from every point in the compass towards the struck one in the most mysterious manner.'

In 1882, the *Eclipse* killed 203 Botleys and the *Windward*, another Peterhead whaler, killed 103. The Norwegian sealers and walrus hunters were eager to get in on the act, and, oddly enough, the Scottish whalers helped to pave the way for them. In 1883, the crew of the *Erik*, under the command of Robert Gray's uncle, Captain Alex Gray, included two or three Norwegians. 'Was it for nothing that these men took the trouble to sail on a Peterhead ship?' asked Dr Gray.

In 1885, using small sloops and schooners, the Norwegians killed 1,300 Bottlenose whales, in 1885 the total rose to 1,700, and by 1891 they had seventy vessels and chalked up a total annual 'kill' of 3,000 whales. The *Eclipse*, however, was keeping its end up. In 1882 it had what Basil Lubbock described as 'a wonderful catch' of Botleys south-west of Jan Meyen. In two months its harpooners fired 338 shots, and 224 hit their target. Fifteen whales were killed and flensed in one day. The Norwegians spread so much destruction among the Bottlenose whales that by 1920 the whalers could only find and kill two or three hundred. At the end of the last war they were back in business. Using fast small-whale hunters, they chased the whales west to the Faroes, Iceland, Greenland and the north-east approached to America.

They took their killing ships to Scottish waters. Farley Mowat the Canadian author of *A Whale for the Killing*, wrote about seeing one of the Norwegian small-whale catchers in Thurso harbour in 1962. The skipper of the seventy-foot vessel, who was also the gunner, described it as 'just a meat shop'. They fitted out at Bergen and went west until they found whales—Minkes, Bottlenose and Potheads (Pilot whales).

They were also in Shetland. When I was there on the trail of the Pilot whale, I met Clem Williamson, who ran a photographic studio in Scalloway. During the war, this small Shetland fishing port was the home of the so-called 'Shetland bus', a base from which Norwegians ran weapons and supplies to Norway under the noses of the Germans. Clem, who was in his late eighties, knew the Norwegians well and didn't much like them. He had never forgotten how they had callously shot a pet seal which used to play around the harbour.

The Norwegians saw large numbers of whales in Shetland waters during the war years. After the war, they came back with their whale catchers, harpoon guns mounted on the deck. They killed off all the whales, said Clem, driving them outside the three-mile limit before harpooning them. He tried to get it stopped, but all that happened was that fishery boats were alerted to keep an eye on them. They were mostly after Minke, but they took Bottlenose and Fin, and I was told later by an old whalerman that they just wanted the blubber, cutting it up, putting it in barrels and transporting it back to Norway.

When long-range whalers from Norway were operating off the Canadian Atlantic coast in the early 1970's they were able to find only a handful of Chaney Johns, and later there were none. Farley Mowat believes that, like the Atlantic grey whale, the Bottlenose will become extinct, despite the fact that the International Whaling Commission gave it provisional protection status in 1977.

Whatever success the Peterhead captains had at whaling, it was their sealing activities that kept them on top while other whaling fleets were going to the wall. By 1845, Peterhead had become David to an English Goliath, the port of Hull, which it displaced as the leading whaling port in Britain. It held this lead until it was outpaced by Dundee in 1872. Throughout these prosperous years, the Blue Toon's success depended as much on sealing as on whaling, and possibly a good deal more.

The sealing statistics for that period are mind-numbing. Records show that in 1822 the *Union* of Peterhead had a catch of five whales—and 2,500 seals. In 1847, the same ship caught two whales—and 7,500 seals. Eelie Bob, the *Mazinthien* sailor who was told to waken up in 1851, must have been bright-eyed and bushy-tailed fifteen years later, for in 1866 the *Mazinthien*, under Captain John Gray, killed 9,300 seals. Yet in later years even the *Mazinthien's* total faded into insignificance, for the largest number of seals taken in one season by a single vessel was 23,000. This would have yielded 270 tons of oil, valued at approximately £14,000. 'Three thousand is not an unusual number to be slaughtered (and flayed) in a single day by a single ship,' said Captain David Gray.

The captain's son, Robert Gray, put these figures against the bloody background of a seal hunt when he described a sealing voyage on the *Eclipse* in 1883. The killing field was among the ice packs off Jan Meyen. Up in the crow's nest, his father spotted the masts of other ships lying far ahead in the ice. The *Eclipse's* engines were working at full speed, black smoke pouring from the funnel, and the ship pushed its way through ice, two feet thick, to join the waiting fleet. When they drew closer they discovered that they were all Norwegian whalers, seventeen of them, lying in two rows with the doomed seals in between—thousands of them, nearly all saddle-backs, old and young, lying close together as if comforting each other. They covered several square miles of ice.

Just before dark the *Eclipse* took up its position opposite the *Geysir*, a rakish-looking Norwegian craft. The evening was fine, but for the seals the sun was setting for the last time. 'Long before the sun rose,' wrote Gray:

> the work of slaughter was in full swing, the old ones being shot and the helpless young merely clubbed. During the day the crack of the rifles, the cries of the seals and the shouts of men were incessant, but in the afternoon as the seals became scarce and the men tired the noise gradually died, and before dark, except for the screaming of gulls, there was silence on the ice, and where there were thousands of happy seals intent on nursing their young there was now only desolation, the ice being streaked with blood and dotted with carcases, or heaps of skins, each surmounted with a flag.

To some, killing seals was simply business, to others it was sport. John George Arbuthnott, who was a passenger on board the *Eclipse* when it sailed to the Greenland seas in 1852 under the command of Captain John Gray, Sen., made the following entries in his diary:

> *April*—I have killed & brought 52 old seals to the ship & would have killed many more but was short of powder. Boats came on board and got 333 old seals. Seals were lying very well.

Monday 3rd May—We have got in all 868 old seals (shot them all). I had glorious sport, having killed 119 seals with my rifle. When I went back for more seals at 5pm I had only 52 balls & shot and killed 44 seals.

Friday 7th—We have now 50 tons in all (blubber) and 670 young seal skins and 2,206 old ones.

The sealing statistics for those years are grim enough, but what it meant in blood and suffering can be seen in the diary entries of Alexander Trotter, surgeon on board the Fraserburgh whaler *Enterprise*, when he took part in his first kill in 1856. He wrote:

This day we got our first seal: the poor creature lay quite still on a piece of ice until our row boat got to her, one of the men shooting her from the boat, and then another running on to the piece of ice and striking her on the head different times with what is called a seal club, although it is rather like a pick than a club. It consists of a long wooden shaft and an iron head having a sharp point and certainly it appears to be well suited for the work it has to perform.

Trotter said the carcase was similar to that of a pig, and not at all like a fish. He was astonished at the immense quantity of blood in a seal—'two full grown oxen, I should say, contain less than one of these creatures no larger than a sheep'.

The days of the Buchan whalers came to an end at the turn of the century—and the years of Arctic slaughter were over. The Greenland Right whale had been virtually wiped out and whalers were turning their eyes to other hunting grounds—to the Antarctic, for instance, which was to become a massive killing ground in the first half of the twentieth century. It was true, as Robert Gray wrote in 1920, that the discovery of new lands meant near-extinction for these 'unhappy animals'. As far back as 1875, David and John Gray had considered taking their ships to the Antarctic, but nothing came of it.

In 1891, both the *Eclipse* and the *Hope* were sold. In 1901, the *Hope* was driven ashore on Byron Island in the Gulf of St Lawrence and became a total loss. The sturdy *Eclipse*, which was perhaps the most famous of all Scottish whalers, went stubbornly on for another half century. She served for a time with the Russian Imperial Navy and after the First World War went back to the Arctic as a supply ship. She sank in 1927, was raised in 1929, and went to Siberia as a research ship. The old veteran finally met her end in 1941, when she was destroyed by a German bomb during an air raid on Archangel.

The last voyage made by Captain John Gray, Sen., was in 1856—he died before reaching home after killing ten whales. His son, John, went into retirement after the *Hope* was sold, but died the following year. Alexander, the least well-known of the three brothers, went into the Hudson's Bay service. With forty-three seasons behind him, Captain David Gray still felt the pull of the sea—he came out of retirement, fitted out the old *Windward* and went whaling again: she was the last whaler to sail from Peterhead. But the man who had harpooned nearly 200 Bowhead whales and 197,000 seals went out with a whimper, not a bang. In 1893 the Greenland seas were bare of whales and the *Windward* brought home only one 'fish'. The voyage took its toll of the old skipper's health and he never sailed again. He died on 16 May 1896, aged 67 years.

CHAPTER FOUR

The Fearsome Night

> But the Lord sent out a great wind into the sea,
> and there was a mighty tempest in the sea,
> so that the ship was like to be broken . . .

The words trailed across the half-deck of the whaler, barely heard over the sound of the wind and the waves. All day it had blown a hurricane. They had passed seven icebergs, rising up in a line, like monstrous sentinels, and now they had gathered for prayer. William Elder, the mate of the *Viewforth*, was reading from the Bible, drawing comfort from the story of another tempest . . . the story of Jonah, who was swallowed by a great fish.

The exhausted seamen, listening to the age-old story of how Jonah had been thrown from the whale's belly on to dry land, must have wondered if they, too, would be saved by a miracle. 'The danger and the sights we have been subject to these last eleven days no pen could write upon,' said the mate. Yet Elder *did* write about the *Viewforth's* ordeal, a stark account of fearsome nights and long days when death was 'within an oar's length'.

Elder's diary described what happened in the winter of 1835 when the Kirkcaldy whaler drifted southwards, 'along an iron-bound coast', trapped in the ice. It is a grim, almost unbearable tale . . . of sick, exhausted men reduced to living skeletons, of a sailor watching his rotting feet fall off from frostbite, of men with their gums hanging loose from their teeth because of scurvy. Through it all, Elder clung to the hope that he had held out to the men in Jonah's story—'that if the mariner's trust was in God, no fears, no dangers of shipwreck, no affliction, would make his spirit fail'.

The narrative is compelling, not only because of its description of the *Viewforth's* torment, but because of what happened when Elder turned away from the whaler and saw with awe-struck eyes the frozen wilderness around him. The killer Arctic could blind you with its terrible beauty. He feared it, yet he found himself looking at it with a kind of wonder. With men sickening and dying around him, he thought that what he saw was 'awfully romantic'. He was attracted to a black cloud 'as dark as midnight' hanging away to the north-west, and he watched a 'beautiful comet star' streaking across the sky, its bright tail glowing in the darkness. Looking towards the land during daylight, he saw huge mountains twisted into 'a thousand fantastic forms', like ruined castles.

It was the darkness—'the fearsome night'—that weighed on him most, but the moon would come up, 'a welcome messenger to cheer us in our darkened paths', and the aurora borealis, that age-old harbinger of disaster, would light up the sky with all the colours of the rainbow, reminding him of a line from Burns—'Or like the borealis race that flits ere you can point its place'.

He watched the skin peel from his men's feet until there was nothing but raw flesh—'the most shocking sight I ever saw,' he said—and he could barely find words to write about it, yet a few hours later he was confiding to his diary his thoughts about the 'beautiful night'.

> The moon is risen. There is not a cloud to be seen in the sky. The last rays of the setting sun are now disappearing (the ice reflecting the light still) for the sun went down four hours ago. Not a hush nor a sound to be heard, for there is no wind. The terrible mountains covered with perpetual snows, the tremendous precipices, the mountains of ice around in all the different shapes you can imagine. The rising moon, the clear blue sky, the still of the evening . . .

He thought it was all 'grand and magnificent', but the sobering reality of his plight returned and as he stood on deck he began to think of home. Tears rolled down his cheeks. He wondered if he would ever see his own fireside again. He wondered, too, if he was prepared to die. 'Sad, sad thoughts,' he wrote, 'my pen is past describing them.' But William Elder survived the nightmare of the *Viewforth*. His ship reached Stromness on 9 February 1836, with fourteen of her crew dead.

Narratives like these contribute much to the history of nineteenth-century whaling. some were the work of captains and officers, but many were written by ship's surgeons, young men who signed up to find adventure and gain experience before settling down to their medical careers. One of them was Arthur Conan Doyle, the creator of Sherlock Homes, who sailed to the Greenland fishery in 1880 on board the Peterhead whaler *Hope*.

Conan Doyle was twenty when he signed on as surgeon of the *Hope*. He seemed to be accident prone, for one day he fell into the sea while watching the men killing and skinning Harp seals. The captain thought he might be safer on the ice and invited him to join him there, but he fell into the water twice more before they day was out. After that the crew called him 'the great northern diver'. In the disastrous season of 1835–36, when the *Viewforth* was trapped in the Arctic ice along with other Scottish and English whalers, a handbill was circulated in Hull telling the story of the 'five hundred seamen brave ice-bound in Davis Straits'. It carried nine verses about their 'Dreadful Hardships and Privations':

> Now 'midst those dreary regions, from their friends far away,
> Hunger and cold they do endure, with one biscuit a day.

There was a touch of William McGonagall about the poem, but the lines were typical of the whaling songs of last century. Some were fanciful, like 'Paddy and the Whale', in which Paddy slipped down the whale's belly and stayed there for six months and five days until, like Jonah, he was coughed up on to dry land:

> And now that he's safely back home on the shore
> He swears that he'll never go whaling no more;
> And the next time he's wishful for Greenland to see
> 'Twill be when the railways run over the sea.

The north-east of Scotland, with its strong bothy ballad tradition, was the source of many whaling songs, which was not surprising, for it was from farms and fishing villages along the Buchan coast that the whaling companies drew many of their recruits. The Buchan folk-song collector, Gavin Greig, was concerned that 'the local whaling minstrelsy' would die out when the Greenland men had gone. Much of it did, particularly the crude choruses that Greig thought were 'rather strong for print'.

There is, interestingly, an old whaling song that is still sung to-day—'Farewell tae Tarwathie', written by George Scrogie, a miller at Fedderate, New Deer, in the early 1850's. This haunting beautiful song was recorded by Judy Collins in a long-playing record called 'Whales and Nightingales'. Behind the voice of the singer can be heard the wailing of the whales, a sad sound, as if they were crying out against their fate. Tarwathie is a farm in the lap of Mormond Hill, near the village of Strichen, and the song tells the story of a lad who left there to seek his fortune at the whaling:

> I'm bound out for Greenland and ready to sail
> In hopes to find riches in hunting the whale.

There are three Tarwathies near Mormond Hill, but no one has ever been able to find out what happened to their whaler boy. He seems, at any rate, to have had a successful trip, for the last stanza mentions the 'bonnie ship' heading home 'bumper-full'. Not everyone had the same luck. Many went off with high hopes and came home vowing never to return to the 'cauld countrie'.

'Highland Laddie' was the tale of another youth who 'went a-whaling' and ended up wishing he was back in bonnie Scotland—'Whaling's not the road to fortune'. 'Highland Laddie' was a favourite towing song among Scottish whalers. In his *Voyage of the Aurora* in 1804, Dr David Moore Lindsay described how, when the *Aurora* was moving up Lancaster Sound, an Eskimo was spotted on the ice singing the song. It was thought that he had heard it from his parents in the old sailing ship days.

There were a number of versions of 'Bonnie Laddie, Hieland Laddie', including one which had a link with Dundee:

> I've shipped for the north on a Dundee whaler,
> Bonnie lassie, Hieland lassie,

Shipped for the north as a whaler sailor,
My bonnie Hieland lassie.
But Greenland's shores are grey and cold,
Bonnie laddie, Hieland laddie,
There's plenty ice but not much gold,
My bonnie Hieland laddie.

There were times when names and places in the songs were changed to suit different area. One traditional song, 'Diamond Ship', had a string of alternative versions. the opening stanza of one went:

The *Diamond* was a ship, brave boys, for Davis Straits was bound,
And the quay it was all garnished with pretty girls around;
Where Captain Gibbons gave command to cross the mountains high,
Where the sun it never sets, brave boys, nor darkness in the sky,

Along the quay of Aberdeen the girlies they do stand,
With their mantles all around them, the tears running down,
Don't weep my pretty fair maids, tho' you be left behind,
For the rose shall grow on Greenland's ice before we change our mind.

The song ended by toasting the health of the *Hercules*, the *Jane*, the *Bon-Accord*, and the *Diamond*, but in another version the ships had become the *Eliza Swan*, the *Battler* of Montrose and the *Resolution* of Peterhead. The *Jane* had its moment of fame in 1810 when it returned to port with the largest cargo of oil ever brought into Aberdeen. It was celebrated in a local ditty:

We'll go into Jean MacKenzie's
And buy a pint o' gin
And drink it on the jetty
When the *Jane* comes in.

Gavin Greig included in his 'Diamond Ship' collection a mysterious verse about a local girl's love for a seaman on the *Jane*—'whose name I daurna tell':

There is ane amang the rest, his name I daurna tell,
For as the sun excels the stars, his beauty doth excel;
His rosy cheeks and cherry lips, his bonnie black rollin' eye,
And for my life I'll be his wife, or for him I will die.

Here's a health to the Captain, another to the *Jane*,
And another to the lad that wears the napkin o' the green.

The pet poets of the whaling fleets cast a wide net in their search for subjects. In 1815, when the North Pier at Aberdeen was damaged by gales, word went round that the harbour entrance was blocked by fallen stone and that the port would be closed during repairs. This meant that ships were unable to leave port, which brought a cry of protest from the whalermen. Out of it came 'The Whalers' Lamentation':

Cease rude Boreas, westering railers,
Leave oh leave our pier alone;
Pity, pity us poor whalers,
We'll ne'er get o'er the bar for stone.

The poem mentioned 'T——d's plans and J——'s pier (T was the engineer Thomas Telford) and the poem damned his plans and improvement. The chorus went:

Davis Straits adieu this season,
Greenland for a year goodbye;
Whales and sharks, you have reason
Chant T——d's praise to the sky.

As it happened, the harbour wasn't blocked. All the ships sailed, which brought an additional verse proclaiming that 'T——d and the Devil can't prevent our going to sea'.

A peculiar feature of 'The Whalers' Lamentation' was that before each verse it carried the names of the Aberdeen whaleships, presumably to show how many vessels were being prevented from going to sea. There were fourteen in all—the *St Andrew, Jane, Neptune* and *Middleton* (Union Company), *Bon-Accord* and *Elizabeth* (Bon-Accord Company), *Diamond* and *Hercules* (Aberdeen Company), *Letiitia, Elbe, Princess of Wales* and *New Middleton* (Greenland Company) and *Dee* and *Don* (Dee Company).

The longest whale song on record was 'Eelie Bob', a name taken from a whaler captain who always tried to be first to get his cargo of oil. It contained a reference to nearly every whaling ship and skipper sailing from Peterhead and is said to have been written by the doctor of the *Mazanthien*, one of the largest and most successful ships of the time. Here again there were different versions, each one carrying the chorus:

Waken up, Eelie Bob, or you're sure to be done
In this year, eighteen hundred and fifty-one;
Waken up, Eelie Bob, or you're sure to be done,
The *Mazanthien's* home with her two hundred ton.

The year 1851 was the first year that the *Mazanthien*, under Captain P. Burnett, sailed with the Peterhead fleet. She had the highest catch of the year, totalling 214 tons from 15,579 seals. The *Columbia*, another whaler named in the song, was also making her first trip from the Buchan port. The *Columbia's* skipper was Captain Robert Birnie—'a very nice man'. Was he the 'Eelie Bob' of the whale song? The *Enterprise*, which had little luck at the whaling and was eventually sold to Fraserburgh, was also in the Peterhead fleet that year.

Gavin Grieg wrote about some of the songs 'cutting up ships, skippers and owners in a somewhat wholesale way'. He thought that the characterisations were too strong for print. He may have had 'Eelie Bob' in mind, for, although it praised Captain Birnie, who 'gives grog to his crew', it also took a backhand swipe at the owners:

There's some of her owners rather windy in their way
They were originally intended for Botany Bay,
But by some of the clever tricks of the hand
They escaped that though they ought to have been hanged.

The crew of the *Enterprise* also came under fire:

There's Wady in the *Enterprise* is next to describe,
With their Windybags and Heather Jock and all their other tribe;
Low, mean in their ways and respected by none,
By all decent people their company is shun.

Some of the whale songs dealt with the loss of men at the Greenland fishing. One, simply called 'Whale Fishing Song', was passed to Greig by a correspondent who, as a boy, heard it sung by the whale fishers in Aberdeen when the last ships were being prepared for the Franklin expedition. The lines went:

> Oh we struck that whale and away she went
> With a flourish in her tail,
> But oh alas we lost one man
> And we did not kill that whale, brave boys,
> And we did not kill that whale.

The fate of the Franklin expedition brought a huge outpouring of verse, including 'Lady Franklin's Lament'. which told of the search for Sir John Franklin and the men of the *Erebus* and the *Terror*. 'Penny of much renown' was mentioned for his part in the search. Whoever wrote this emotionally over-charged piece was no poet, but tears were shed by many who read it:

> They've sailed east and they've sailed west,
> Round Greenland's coasts they knew their best
> In hardships they vainly strove
> On mountains of ice their ships were drove.
>
> In Baffin's Bay where the whale-fish blows
> The fate of Franklin there's no one knows,
> Which causes many a wife and poor child for to mourn,
> And sad forebodings for their return.

The thread that ran through many of the old whale songs was despair; despair at what men found when they reached the 'fishery'—'Greenland's dark and dangerous place, and there grows naething green'—and despair at the death of the 'brave boys' who went a-whaling—'Crying alas for my four pretty men, the darlings of our crew'.

Not all the whaling ditties were gloomy. One song, 'Peterhead', said that in the whaling days 'the lasses sang from morn till eve' when their men went to the Greenland seas. When they came home, money flowed and no one had a care. But the whaling days and their fleeting prosperity finally came to an end and the line fishers and drifters took over. The verse went:

> But brave hearts never did give in,
> Whatever come and go,
> And Peterhead will flourish still
> As in the days of long ago.

To-day, as another kind of oil boom makes it self felt in the Buchan port, that poetic prophecy has come true.

CHAPTER FIVE

Bessie's Blessing

Up on Brinkie's Brae I felt the wind on my face and wondered if Bessie Millie was still practising her black art. Looking down over the rooftops of Stromness to the harbour, watching the stir of traffic in the bay, I was thinking of the ships that sailed out of this Orkney port nearly two centuries ago, past the Point of Ness and into the swirling tides of Hoy Sound. They went on their way with Bessie's Blessing; some with her curse. She lived in a house at the top of the Brae, and they say she was a witch who sold favouring winds to gullible sailors.

The whalemen who fought gales and storms in the Arctic must have felt bitter about the crone who traded false hopes for a handful of coins, but the fair winds that Bessie sold blew a breath of life into Stromness. It was in Bessie's time that the town grew in importance, becoming a last port-of-call for west bound ships, not only for merchant vessels, but for Hudson Bay ships heading for the 'Nor-Wast' and whalers bound for the Davis Straits. By 1750, the cluster of houses at the foot of the bare lump of a hill at Hamnavoe had begun to expand—up Brinkie's Brae.

It is a quaint, old-fashioned place. Eric Linklater called it 'eccentric'. Its main street zig-zags through the town like a drunken sailor, clutching the edge of the shore. The street is fed by narrow lanes and closes running down from Brinkie's Brae, including Puffer's Close, named after a town crier, the Khyber Pass (once called Garrioch's Close) and the curious Hellihole. Rough, muddy and dusty at the beginning of last century, now neatly flag-stoned, the street had a maximum width of 12 feet, narrowing to 4 feet at Porteous's Brae. When he visited Stromness in 1814, Sir Walter Scott complained that you couldn't get a cart, or even a horse, down it. It is still tight in places, but now you have to dodge cars, not horses.

The men who went to the Arctic last century left an imprint on Stromness that the years have never rubbed out. There are reminders of the whaling days all along this meandering thoroughfare . . . the Pier Arts Centre, which was where the Hudson Bay Company recruited its men, the house built by Mrs Christian Robertson, a shipping agent who in one year engaged 800 men for the whaling, the cannon on Stranger's Brae which signalled the arrival of Hudson Bay Company ships, the house where Lady Franklin lived when they searched the Arctic wastes for her missing husband.

There are no back gardens in this street of memories; instead, there are piers and jetties. The householders have the salty taste of the sea on their lips and Bessie Millie's fair winds whistling under their doors. They look out on a harbour that has seen Captain Cook's *Resolution* and *Discovery*, and Sir John Franklin's Arctic-bound ships, and the Stewarts of Massater. The vessels all watered at Login's Well, which was sealed up some sixty years ago, but it has an inscription on it listing some of its famous visitors. There was stronger drink across the road at Login's Inn.

The street is divided into sections, each with a different name—Victoria Street, Dundas Street, Graham Place—but I can't help thinking that Orkney's favourite son, George Mackay Brown, had a point when he said that the whole lot should be called the Planestones, the name that was originally given to one section of it. The Orkney poet lives near Stromness Museum, an old building musty with the smell of history. There is so much sea lore in it that you almost find yourself walking with a sailor's gait when you pass through its doors. It may have something to do with the poem that hangs on one of the walls, telling how a man's thoughts would often 'taak a spang aff tae the wild Nor'-wast'.

On winter nights I whiles can feel
Me cottage gaan adrift,
An' wance again I grip the wheel
Tae the sea swaal's aisy lift.

Robert Rendall, Orcadian poet, artist and archaeologist was writing about an old sailor's dream of the days when he was 'ruggan afore the mast'. Not far from the Stromness Museum, in his home in Rae's Close, Captain John Gray must sometimes feel 'the sea swaal's aisy lift' as he listens to the muttering tides almost on his doorstep. This old seafarer, who once sailed on supply ships to the South Antarctic whaling grounds, was born when Arctic whaling was coming to an end. He remembers seeing Caa'ing whales and he remembers his grandfather talking about them being driven ashore at Stronsay and Westray.

John, born on the island of Papa Westray, says he was 'brought up with the sea'. His father was a crofter-fisherman and they lived mostly on small fish—salted cuithes—which was about the only food you could get in those days. 'The winter eating,' John called it. There were so many of these tiny fish that you could scarcely see the bottom of the sea for them. He left home when he was seventeen and got a job as an ordinary seaman on a Leith ship sailing to the Continent, and in 1928 he went to the whaling.

He was recording fiddle music when I knocked on his door, for he is a great fiddle enthusiast. Most of the whalers had an Orkney or Shetland fiddler on board, men like Peter Hunter of Bressay, who died on board the whaler *Swan* in the Davis Straits in 1836. When I sat in John's house and listened to his tales about voyaging to the Antarctic, I felt like the old sea dog in Robert Rendall's poem, who could hear 'the very timmers o' the roof creak as we dunt along'. He spoke about how they carried supplies to Tristan da Cunha on their way to the Antarctic; about life in South Georgia, where the whaling stations had names like Stromness and Leith; about his trips on the *Seringa*, carrying supplies—mostly coal—for the whale catchers; and about the whalers—they were all young men—who tried to change themselves from beardless boys to grown men by letting their hair grow. 'They grew whiskers down to here, you know,' said John, pointing to his waistline.

He remembered how some of them had wanted whales' eardrums to take home. 'They would go in among the blood and guts and cut off the eardrum,' said John, 'then you could boil the flesh off it and if you had it lying on your sideboard you listened to it.' John thought that the sounds they heard might have been whales whistling to each other. Many of the whalermen passed the time during the long voyages by doing scrimshaw work on whalebone. There is a decorated whale's skull in Stromness Museum, and in an antique shop in the Plainstones I saw a whale's tooth with a sailing ship carved out of it. It cost £175. Nowadays, artificial whales' teeth have replaced the real thing.

There were twenty whale catchers in South Georgia in those days. At one time, they called them whale killers, but to mollify a sensitive public the name was changed to 'catchers', which was hardly an accurate description of the bloody work they did. The men on the whalers were Norwegians, although Shetlanders were employed at a later date. Mostly, however, the island workers were on the mother ships or working ashore at the whaling stations.

The whalers fussed around the mother ships 'like a lot of kittens', hauling their catches to the bigger vessels so that they could be flensed and cut into sections to go into steam boilers. John once saw a catcher hauling fifteen whales at one time, towed by the tail, side by side and he remembers a Shetland captain using whale carcases as fenders when the catchers came alongside his ship. Generally, they were after the Blue whale or the Fin whale. There was one whale captain who caught 365 Fin whales in a season. 'Those whales could have been anything up from 90 to 100 feet long,' said John. 'A ton a foot—90 feet and they would be 90 tons. Just like a little ship.' It was the slaughter of the Blue and Fin whales that brought the first moves by the world's whaling nations to conserve the dwindling stock of whales.

John was on the Antarctic run for four years. He got his master's ticket in 1931, and he was still sailing as a master in 1973, when he was seventy, doing pilot work in Scapa Flow and carrying out contracts for the oil companies. 'It was a great life', he says. Now he shares his memories with other old sea dogs at the Pierhead Parliament in Stromness, looking out over a bay where the whalers once dropped anchor and Sir John Franklin's ships put into port on their way north to find the elusive North-West Passage.

There is a road in Stromness named after Franklin, but George Mackay Brown once said it should be called Rae Road, for the real hero of the tragedy that wiped out the Franklin expedition was a Stromness man. Instead, Dr John Rae gave his name to Rae's Close, where John Gray lives. Rae, chief trader with the Hudson Bay Company, played a major role in mapping out northern Canada and was also the first man to discover what happened to the crews of the *Erebus* and *Terror*. He was a character out of fiction, lean, tireless, good-looking, with deep-set eyes and black side-whiskers, said by R. M. Ballantyne, the author, to be a man 'full of animal spirits'.

He had a solitary Orkney childhood, often going off on lonely walks with his Newfoundland dog, and he spent his time boating, fishing and shooting with an old flintlock. They were skills that stood him in good stead when he tramped across the Arctic wilderness. When he was asked to join the search for Franklin in 1847 he had just successfully charted 625 miles of coastline, travelling 1,200 miles on foot, living off the land.

He was a loner. His relationship with his Navy colleagues on the Franklin search was an uneasy one. They could never understand a man who appeared to be as primitive as the natives, who copied the Eskimo way of life, adopted their dress, ate their food, lived in snow houses, slept under caribou blankets, travelled by dogteam, and even shot his own meat, something that other leaders would never think of doing. It was said that he could supply his whole party with venison without troubling himself too much.

The 1847–48 expedition shed no new light on the fate of *Erebus* and *Terror* crews and the search dragged on long after hope of finding anyone alive had been abandoned. In April,

1854, while surveying the Boothia Peninsula, Rae found the first key to the Franklin mystery—and put himself in line for an award worth £10,000. At Pelly Bay he met an Eskimo, Innook-po-zhee-jook, who said he had heard stories from other natives of thirty-five or forty white men who had starved to death some years earlier, about twelve days' journey away.

Later that year, it was established that the bodies had been found near the estuary of the Great Fish River. The Eskimos brought a mass of relics to Rae at Repulse Bay—one of Franklin's decorations, a small plate with his name on it, silver forks and spoons, a surgeon's knife, a gold watch, and other items. They also told Rae that Franklin's starving men had committed acts of cannibalism. When this news reached Britain the reaction was shock and disbelief. The writer Charles Dickens, while obviously believing that the 'treacherous and cruel' Eskimos might eat each other, thought it was 'in the highest degree improbable' that Englishmen would eat Englishmen. Doubt was cast on both Rae's discovery and on the cannibalism report, but the Orkney explorer held his ground. He got his £10,000 with £2,000 of it going to his men.

The house where John Rae spent his childhood can still be seen in Stromness's winding main street, next to the Northern Lighthouse depot. The building, known as the Haven, was at one time used by Hudson Bay Company agents, and Rae's father lived there from 1819 to 1836. In 1851, Lady Jane Franklin and her niece Sophia stayed at the Haven while visiting Stromness during the hunt for her missing husband.

She went to see the house where Sir John had stayed before sailing for the Arctic, and she took cake and brandy with John Rae's 75-year-old mother—'the most beautiful old lady we had ever beheld', wrote Sophia. At that time Lady Franklin had a high regard for the Orkneyman, but her admiration diminished after she had heard his report about members of the Franklin expedition engaging in cannibalism. The islanders were intrigued by the two Englishwomen, for they remembered seeing Sir John Franklin when the *Erebus* and *Terror* called at Stromness in 1819.

The whalermen who put in to Stromness knew all about the Franklin drama, for many of the whale ships had taken part in early searches for the missing ships. In 1848, the Admiralty offered a reward of 100 guineas to any whale ship which gave any authentic information about the *Erebus* and *Terror* in Lancaster Sound, and an offer from Lady Franklin was published in the whaling ports—£1,000 for any ship that found the expedition and another £1,000 for any vessel which made 'extraordinary exertions' to find Franklin and his party and bring them back to England.

Stromness was a busy port when Sit John Franklin went off to find the North-West Passage. Hamnavoe was full of ships, some taking men off to the 'Nor-Wast' to work for the Hudson Bay Company, others bound for the Davis Straits and the whaling grounds. As well as Hamnavoe, the ships used bays at Longhope and Widewall in the south and Deer Sound, where a ship called the *Fame* was destroyed by fire in 1823. The *Aberdeen Chronicle* reported in 1816 that there were 34 whaling ships at Stromness, 25 or them from Hull. In 1841, when whaling was on the decline, the Rev Peter Learmonth, a Stromness minister, said that for the past seven years and average of 295 men annually had gone off to the whaling, and that before that the number had been much greater.

Orkney came into its own in the whaling business after the fishing switched from Jan Meyen and the Greenland waters to the Davis Straits. Before that whalers picked up their stores and crews at Lerwick. Shetland was said to be a place of wild winds and wild men, and Lerwick, with its narrow, refuse-strewn streets and ill-fashioned shops, its whisky-dealers and prostitutes, had a murky reputation. Stromness, according to one account, was not much better, 'an irregular assemblage of dirty huts' and 'scarcely anything deserving the name of a street' in the place.

Stromness not only became the main jumping-off point for whalers from Scotland and England; it also had its own whaling fleet—one vessel. In 1813, the Orkney Whaling Fishing Company was formed in Kirkwall and the whaleship *Ellen* bought from Leith. In 1822 the business was sold and a New Orkney Whale Fishing Company was formed, with the *Ellen* operating from Stromness. It went out of existence in the late 1820's and its 'oily house' in Kirkwall was used for boiling blubber from Pilot or Caa'ing whales.

The whaling companies had as their agent in Stromness a local merchant, Mrs Christian Robertson, who had little trouble providing the ships with the sort of men they wanted. In one year, 1,400 men applied for jobs on the whale ships—800 were engaged, 600 turned away. 'The men are very plentiful here,' she wrote to one company manager. Her house, two buildings turned into one and known as the Double House, can still be seen, gable-end to the sea, at the south end of the town.

Christian Robertson's recruits went off to the Davis Straits with high hopes. they boasted in their shanties that they would come home 'wi' a ship full o' oil', but there were times when they returned with empty holds—or, much worse, with broken ships and broken men. Stromness wept at the sight of survivors coming ashore from whalers battered and crushed in the Arctic ice. 'It drew tears from the eyes of many an unconcerned spectator,' said one report, when the *Lady Jane* of Newcastle arrived in port. 'Such a heart-rendering scene Stromness never witnessed before.' There were twenty-five Orcadians on the *Lady Jane*—thirteen died.

The bad years ran like a scar through the history of whaling. The most disastrous year was 1830, when out of 91 British ships in the Davis Straits 19 were lost and 21 returned clean. Two of Peterhead's 13 ships—the *Resolution* and the *Hope*—were wrecked. Dundee lost two ships, the *Achilles* and the *Three Brothers*, and the *Baffin* and *Rattler* of Leith were wrecked, as was the *John* of Greenock. Four of Aberdeen's ten ships were also wrecked—the *Alexander*, *Laetitia*, *Princess of Wales* and the *Middleton*, whose successors, *Middleton II*, was to meet the same fate a few years later. Nearly every ship was damaged.

Stromness became a mercy port, sharing the agony of the whalermen. When a large number of ships were reported missing from their home ports in 1835–36, the Admiralty commissioned the *Cove*, under Sir James Clark Ross, to go to their aid, but the vessel was forced back to Stromness by bad weather. It turned out to be an unexpected stroke of luck, for she was there when the stricken ships arrived in port. Two of them, the *Jane* and the *Viewforth* of Kirkcaldy, had on board 34 men 'in a dreadful state of suffering'. Nineteen of them were part of the crew of the *Middleton II*, lost in the ice. Ross, working with James Login, set up a temporary hospital in a large empty house and put his ship's surgeon in charge of it. Supplies and medicine came from the *Cove*. The house, owned by a Mrs Humphrey, held about 28 patients, but her friends and acquaintances stepped in with offers of help.

The *Viewforth* arrived in Stromness on 14 February 1836. There were fourteen dead men on board, and only seven men were able to work the vessel. One of its officers was William Elder, a deeply religious man, who said they had 'escaped from the very jaws of destruction'. 'O Lord, what can we expect of thy hand?' he wrote in his journal. It was a cry that was to be heard again in 1837, when the ice closed in on six ships trapped in an impenetrable field of ice in Baffin Bay. Among them were the *Advice* and the *Thomas* of Dundee, and the *Dee* of Aberdeen.

CHAPTER SIX

Breaking-up Yard

From the Kirk on the Hill you can look south to Scapa Flow and across the Sound of Hoy to the vast emptiness of the Atlantic, or north to the lochs that lie like giant dewdrops across the face of Orkney. Under your feet is Harray, known at one time as the parish of a hundred lairds. This is where Adam Flett, an Orkney whalerman, is buried, in the old graveyard of St Michael's Church, where the rich Orkney countryside dips away to the sea. Not far from Adam's grave is the burial place of James Tulloch, his wife's cousin, whose body was brought home from the Arctic in a coffin. The church stands on a little hill overlooking Loch Harray, with the whale-back outline of Hoy in the distance. But Adam's tombstone faces away from Hoy, almost as if he had turned his back on an island that held only bad memories for him.

Harray is the only landlocked parish in Orkney. Despite this, Adam was drawn to the sea as irresistibly as if he had been born on the water's edge, for all Orcadians have the sea in their blood. His birthplace was at Lynmath, not far from Corrigal Farm Museum, where one of his descendants, Harry Flett, is curator. It was in the museum that Harry showed me a huge Bible carrying the inscription, 'Robert Flett, Lynmath, Harray, June 4, 1847'. Robert Flett was Adam's brother, one of eight children—six of them sons—born to James and Margaret Flett.

Not long after I had been to Lynmath and the Kirk on the Hill I walked along a narrow path that leads from the Outertown of Stromness to another ancient graveyard, which also faces across the water to Hoy. From this cemetery, whose faded tombstones stretch almost down to the sea, the path follows the coastline to the Black Craig cliffs, which rise to a height of some 360 feet above the shore. George Mackay Brown once said that he was rather afraid of the Black Crag, as he called it. If he slipped, he wouldn't stop falling until he was in the Atlantic with the seals and lobsters. When he was a boy there was a saying, 'The Black Crag claims a victim every seven years', although he himself never heard of anyone going over it during his lifetime.

But it was another boy I was thinking of when I picked my way along that rocky stretch of coast. When young Adam Flett left school he went to work at the Black Craig quarries. From there he could see the whalers punching their way up the Sound of Hoy and past the Kirk Rocks, heading north to the Davis Straits. Dreams of adventure and a quick fortune must have stirred his imagination, for when he was only seventeen he gave up quarrying and went to the whaling. In 1836, after several voyages to Greenland, he joined the Aberdeen whaler *Dee*. It was the last time he ever went to the Arctic whaling . . . .

The story of Adam Flett mirrors the lives of all the Orkneymen who went to the Arctic fisheries, lured by the promise of riches and tempted by the thought of freedom from the monotony and poverty of life in the northern isles a century ago. Instead, many found death and disease. Some came back with stumps instead of hands and feet, their fingers and limbs crudely sawn off by ill-equipped and inexperienced ship's surgeons when the whalermen became victims of frostbite. The graveyard of their hopes was Melville Bay, a fearsome stretch of water in the Davis Straits, where dozens of ships were destroyed in the ice. The whalermen called it the 'Breaking-Up Yard', and when they sailed through it they kept a bundle of clothing close at hand in case they had to abandon their vessels.

Many of the Arctic whale-hunters were Fletts from Harray—'Harray crabs', their neighbours called them, mocking their land-locked isolation from the sea, and no doubt some of the taunts came from Hoy, whose crofter-fishermen were long-standing rivals of the Harray folk. Janet Sinclair, who lives near Kirk on the Hill, has been studying the Harray families for a number of years. She found that in 1841, only a few years after Adam Flett sailed on the *Dee*, there were 772 people in Harray, 332 of males and 440 females. She discovered that the men who went to the Davis Straits were marked down in official records as Straitsmen.

The *Dee* left Stromness on 9 April 1836, with the Straitsman from Harray, Adam Flett, on board. She made the ice on 15 May, and three months later they were in Pond's Bay, where they saw their first whales. They picked up three dead whales and killed another three. By the end of August most of the whalers were heading for home, but six lingered on in Baffin Bay. The *Dee* was one of them, along with the *Advice* and the *Thomas* of Dundee. The others were the *Norfolk* of Berwick, the *Grenville Bay* of Shields, and the *Swan* of Hull. By the first week of September all six vessels were trapped in an impenetrable field of ice.

Back home, as the months passed, people remembered the disasters of 1835 and waited anxiously for word of the missing ships. They remembered, too, how nineteen whaleships had been crushed in Melville Bay in 1830, when more than a thousand shipwrecked whalermen had taken to the ice. They lived in tents or underneath whaleboats, raiding the liquor casks of stricken vessels and engaging in an orgy of drink that became known as the Baffin Fair. The scene took on a grim carnival-like atmosphere from the red flames of burning ships, for vessels that foundered in the iced were always set on fire after everything had been salvaged from them.

Only about ten men died in the Baffin Fair episode and some were drunken seamen who wandered away from the tents. Seven or eight times as many died in 1837, despite official efforts to save the whalers. In January of that year, the Government offered a £300 bounty to any relief ship prepared to sail for Baffin Bay before 5 February, and a further bounty of £500 to any vessel able to help the trapped whalers. There was a third bounty of £1,000 for any ship that actually rescued a whaler from the ice. The *Traveller* of Peterhead and the *Princess Charlotte* and *Horn* of Dundee were among the vessels that set off on this mercy mission.

A whale is flensed on the after-deck of a factory ship. On the upper jaw can be seen the baleen with which the whale sifts its diet of krill from the sea.

The scene on board a Salvesen factory ship as cutters work on the carcases of the whales.

Two Salvesen whale-catchers in South Georgian waters. The leading catcher is the *Southern Truce.*

The *Southern Harvester*, one of the Salvesen factory ships that operated from South Georgia.

A harpoon gun on board a whaler in Shetland. The invention of the bomb harpoon by the Norwegian whaling master Sven Foyn revolutionised the whaling industry (*Picture by courtesy of Shetland Museum*).

Ready for flensing. The scene at a whaling station in South Georgia (*Picture by courtesy of Shetland Museum*).

A cutter begins work on the flensing of a dead whale at the Alexandra Whaling Company's whaling station at Collafirth in Shetland (*Picture by courtesy of Shetland Museum*).

A dead Polar bear is hauled aborad a whaling ship. Polar bears were shot for their furs or simply for sport. Some were sold to museums, while others were brought back alive to be sold to British zoos.

The carcases of six whales tied up behind the *Southern Harvester*. A whale catcher lies a short distance from its parent ship.

An unusual view of a harpooner firing the harpoon gun on the deck of a South Georgian whaler.

Captain David Gray, master of the Peterhead whaler *Eclipse*, and his son and chief mate, Dr Robert W. Gray (*Picture by courtesy of Arbuthnott Museum, Peterhead*).

Josie and the whale, not Jonah! Josie Manson (left), from Olna in Shetland, is seen sitting in the jaws of a whale at a whaling station in South Georgia. This one-time whalerman scoffs at the story of Jonah and the Whale.

Bulky oil-supply ships nudge each other at Porca Quay, Aberdeen. Whale ships once set sail from here to make their money in another oil boom—the hunt for whale oil. Near Porca Quay was the boiling house. The citizens of Aberdeen found the smell of boiling blubber 'quite disgusting'.

An abandoned whale catcher lies half under the water alongside a jetty at Grytviken in South Georgia. Her harpoon gun can still be seen mounted on her deck (*Picture by Graham Page*).

Captain John Murray in his wheel-house.

Austin Murray, Captain John Murray's son, is seen with a pair of walrus tusks brought home by his father. The Eskimos engraved them with kayaks, reindeer and dog sleds.

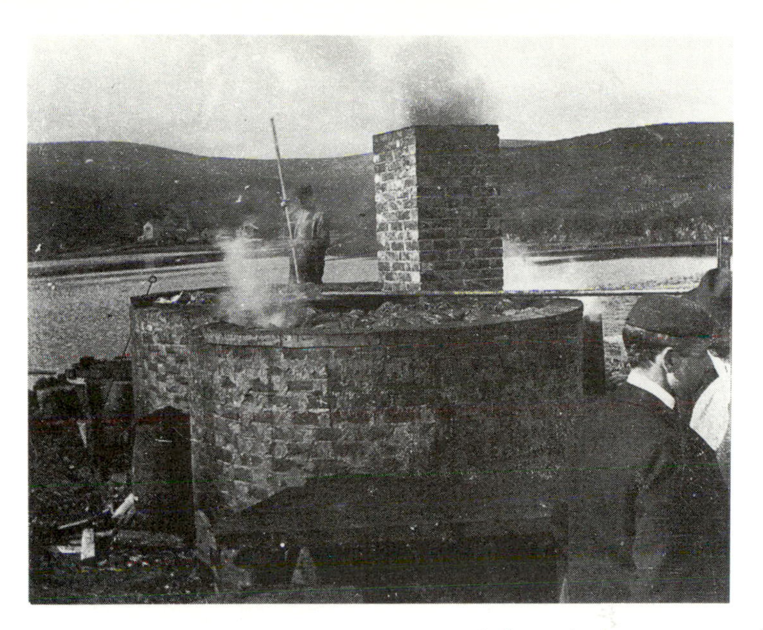

Blubber is boiled in the try pots at Norrona whale station at Ronas Voe in Shetland. The year was 1904. Shetland fishermen complained about the pollution and the effect on their fishing (*Picture by courtesy of Shetland Museum*).

Ecclesgreig Castle, near Montrose. It was from these castle gardens that the laird's son, young Osbert Clare Forsyth-Grant, watched whalers from Montrose set out on their long journey north to the Arctic. He made up his mind that one day he would follow them—an ambition that was to cost him his life.

A striking portrait of Captain Alexander Murray, father of John and his brother Alexander, and grandfather of Austin Murray. All three Murrays served on the whaler *Perseverance* and all commanded her at different times.

William Wilson, from Graemsay in Orkney, went to Stromness to become a whaler and sailed for the Davis Straits on the *Emma* of Hull. The following year he signed up on the *Wildfire* and was wrecked when the whaler became ice-bound. His next ship was the *Intrepid* of Dundee. She left Stromness in March, 1870. The last whaler to ship part of her crew from Stromness. When Wilson died in 1922 he was the oldest fisherman in the town (*Picture from Stromness Museum Collection*).

Captain William Adams, the well-known Dundee whaling master, is seen with his telescope in the crow's nest of his ship (*Picture by courtesy of Dundee Art Galleries and Museums*).

Eenoolooapik, the Eskimo brought back to Aberdeen by Captain William Penny. He dressed like a gentleman, showed perfect manners, and won over the Scots; but when he was asked if he would ever return to Scotland he shook his head and said, 'Too much cough' (*Picture by courtesy of Marischal Museum, Aberdeen*).

Urio Etwango, an Eskimo brought home from the Arctic by the whaler *Maud*, is seen on board the vessel in Lerwick harbour in March, 1887 (*Picture by courtesy of Shetland Museum*).

A modern generation of youngsters learn sea-craft at Stromness, from where whalers set sail for the icy Davis Straits last century. It was here that they picked up their crews and loaded their supplies.

The grave of Adam Flett of Nistaben in Harray, Orkney. He was a member of the crew of the Aberdeen whaler *Dee*, which returned from Baffin bay with 46 dead men on board. Flett was one of the survivors.

Josie Manson on his fish farm at Olna, Shetland. It stands on the site of an old Norwegian whaling station.

Football on the ice was a favourite sport among Scots whalers. The picture shows a Dundee team lined up for the camera. Polar bears sometimes interrupted play, sending the fitba'-daft whale hunters racing for the safety of their ship (*Picture by courtesy of Dundee Art Galleries and Museums*).

A pair of pipe-smoking Eskimos—Terraguaire (left) and Basia—photographed by Captain John Murray.

Three smart Eskimo girls in East Greenland. The girl in the centre may have been taught to play the melodeon by a Scots whalerman, for the melodeon was a favourite instrument with the Eskimos, who loved music. They has a particular fondness for Scots songs.

Arctic whaler, James Milne, a native of Pennan in Aberdeenshire, who married and settled in Kirkwall, Orkney (*Picture from Stromness Museum Collection*).

The crew of a Peterhead whaler. These were the men who wet whaling more than a century ago, facing deadly Arctic conditions in the hope of making their fortunes. As the picture shows, some were mere boys (*Picture by courtesy of Arbuthnott Museum, Peterhead*).

Nouyabik, an Eskimo brought to Peterhead from Baffin Land in 1925. He stayed for a year and returned home in the Peterhead whaler *Albert*. The highlight of Nouyabik's visit was a drive in an 18 h.p. Star car (*Picture by courtesy of Marischal Museum, Aberdeen*).

The whalers in action—a shot taken during the 1957–58 Antarctic whaling season. A catcher is delivering dead whales to the factory ship, while in the foreground the loaded harpoon gun of another catcher can be seen.

Looking across to Hoy from a croft near Stromness. Whalers sailed up the Sound of Hoy on their way to the Davis Straits.

An Eskimo sits beside the 'kill' from a walrus hunt. Walruses were shot for their hides and tusks. The *Balaena* killed 600 of these sea-horses along the shores of Franz Josef Land. Walrus hides sold at 1/6 per pound to the makers of bicycles, and tusks fetched 2.6 per pound.

The legacy of the Antarctic whalers. This is how the abandoned South Georgia whaling station at Grytviken looks to-day, littered with debris, shattered buildings and derelict machinery. In the early years of this century a different kind of litter lay on the shore—the ribs, jaws and huge skulls of dead whales (*Picture by Graham Page*).

A common practice among Antarctic whalers was to use dead whales as fenders between ships. Here, a 'fender whale' is seen between a small vessel used to transport whale meat from the factory ship to a refrigerator ship.

The whalers, tiny specks in the middle of an immense sea of ice, kept as close to each other as possible. Only the *Swan* became separated from the group, drifting so far northwards that no one knew what had happened to her. The *Dee* dwarfed by huge icebergs and tossed about by the movement of the ice, reeled from side to side like a drunken sailor, each roll of the vessel marked by a terrifying crash of ice and water—'like the convulsive groans of an earthquake', said a survivor. The men on the Aberdeen whaler lived on the knife-edge of disaster as the ice opened and shut on their vessel like a steel trap, squeezing and crushing the ship.

The first victim was the *Thomas*. On 12 November she was listing so badly that her crew had to crawl over the deck on their hands and knees. Next morning she was a total wreck and two of her crew were dead. About sixty men from the *Dee* and the *Advice* spent three days recovering provisions from the stricken ship. The food helped to eke out their meagre rations, but it brought only temporary relief.

From early September, the *Dee's* master, Captain Gamblin, imposed strict rationing. They saw bears and foxes near the ship but were unable to shoot them, and when the hunger pains became too bad some men cut up the tails of the whales that he been caught earlier in the voyage. They left the pieces to bleach on the ice, then fried them in a pan with a little fat. Others refused to eat the meat. David Gibb, a member of the *Dee's* crew, who later wrote an account of the whaler's voyage, said that eating whale meat was 'exeptionable', but the end justified the means.

In fact, it was regarded by many whalermen, particularly the officers, as a loathsome thing to do. 'Under the cravings of hunger,' wrote William Elder, of the Kirkcaldy whaler *Viewfort*, in 1835, 'man is quite another being and will greedily devour what he would have before counted poison.' When the *Viewforth* ran out of food, Elder wrote about seeing 'a scene that baffles description'—starving men eating blubber normally used for making oil for the ship's lamps. 'The very smell was enough to sicken one,' he declared. 'It shows plainly that when a human being has not the means of subsistence he throws off his proper nature and assumes another more savage and desperate.'

The *Dee*'s crew had no such qualms. David Gibb said that cooked whales' tails made 'a tolerably agreeable meal', adding that very few were fastidious in their taste. Those who did eat it benefited from it, but their feeling of well-being was short-lived. 'The death monster scurvy', as Gibb called it, began to take its toll. By mid-November, twenty-one men were ill with the disease, their gums swollen, their mouths gripped by a deadening pain, making eating a terrible agony.

The Old Year died in darkness and despair. The wind, wrote Gibb, screamed through the frozen shrouds as if a death whistle had been sounded for the men on the *Dee*. For many it had. The first man died on 11 January. He was James Corrigal, from Orkney, who was in his fiftieth year. He died from scurvy, and his body was wound in a blanket and dropped through an opening in the ice. Eight days later, William Besley, from Aberdeen, was dead, and on the 27th another Aberdeen man, Andrew Bennet, also died.

The crew grew weaker. When the mate wanted to take in two reefs of the topsails, only fifteen men were able to go aloft. The worst blow came on 3 February, when Captain Gamblin died. He had been ill for a number of weeks, his body wasting away day by day. His powerful voice was reduced to a child-like whisper. The carpenter of the *Dee* was too ill to make a coffin for his dead skipper, so the carpenter of the *Grenville Bay* crossed the ice and did it for him. Afterwards, the coffin and its corpse were laid out on the quarter-deck.

Nature provided its own dirge. It came in the shape of a howling gale, beating against the whaler as it drifted south, its dead master lying on the deck. Many of the men had taken to their beds, and only eight or nine were able to do duty. The frost increased. Water in the casks became solid ice and every chest between decks was white with frost. Even the blankets on the beds were covered with ice.

Now the sick and dying men faced a new scourge—vermin. The beds swarmed with lice, which lodged under their skin and fed on their flesh. They could do little about it. On top of that, most the sick men suffered from violent attacks of diarrhoea. The *Dee* had become a death ship. Almost every day another corpse was dropped through a hole in the ice. One victim was an apprentice boy called Alex Reid, who died in the arms of David Gibb.

The first few days of March saw four more deaths. Now, when it looked as if they would break clear of the ice, there were only six men who were still fit for duty. One of them was Adam Flett. The mate of the *Dee* sent over to the *Grenville Bay* to ask if they could give any assistance when they reached the open sea, but there were twenty men sick on the Shields whaleship and they were unable to help. The *Norfolk* was about seven miles away, and the *Advice* twice that distance.

On 15 March the body of the Aberdeen man, William Stirling, was buried—the last corpse to be sewn up in a blanket and put down through the ice. Five months and eight days had passed since the *Dee* first became beset in the ice, and in that time it had drifted south for nearly 700 miles. More men were to die before the ship reached Orkney. Among them was James Tulloch, the cousin of Adam Flett's wife, who died of scurvy on 3 April.

The *Dee*, the *Grenville Bay* and the *Norfolk* broke clear of the ice on 16 March, followed by the *Advice* on 17 March. The *Norfolk* lost eight men, but of all the stricken ships she was the healthiest—and the luckiest. On the second last day of the month she fell in with the *Lord Gambier*, one of the rescue ships, and was given fresh provisions. The *Grenville Bay* got assistance from no fewer than eight ships. The *Advice*, which was missing until June, finally turned up in Sligo, with only seven men out of forty-nine still alive, including ten from the *Thomas*. The ship was said to be a 'horrible spectacle'.

The first signs of relief for the battered *Dee* and its handful of survivors came on 20 April, when a brig was sighted standing to the north. A distress signal was raised, but to the crew's horror the men on the brig appeared to ignore the signal. The ship vanished from sight. Five days later a fishing boat was spotted, but again the *Dee's* hopes were shattered when the boat closed with the whaler and called up to them. One of the *Dee's* crew later described what happened after that:

> This morning at six o'clock we saw a boat containing eight men; we hove to, close to the windward of the boat; we waved upon them to come alongside; they came and asked us in English what we wanted. We told them that we had been all winter in Davis' Straits; and that, if they would come on board, and assist us to Stromness, they would be well paid for it. They would not come on board. We hove two pieces of meat into their boat; the boat being half loaded with fish we hove down a rope to them and asked of them to give us a fish, but they refused, and pushed off.

At six o'clock that evening the *Washington*, a barque from Dundee, bound for New York, saw the Aberdeen ship in distress and changed course towards it. The *Dee's* men discovered that they were near the Butt of Lewis. Stromness lay almost within sight, yet they doubted if they could reach port. When the *Washington's* master was told that only three of the crew were able to go aloft, he sent four men on board to help, and then climbed aboard the whaler himself with wine and food. The stricken ship was towed to the Sound of Hoy, where a pilot joined them. Later that evening, a surgeon appointed by the Government boarded the ship to tend to the sick, and at eleven o'clock the *Dee* finally anchored in Stromness harbour, its long ordeal over. Forty-six men had died on board the whaler, nine of them belonging to the *Thomas* of Dundee.

The owners of the *Dee* sent a shipmaster from Aberdeen to take the vessel back to its home port. It arrived in Aberdeen Bay on the morning of 5 May and entered the harbour at noon. The quay was crowded with people, among them relatives of the men who had died. Even though they had been told the names of the dead, they refused to believe it until they saw for themselves. 'Weeping widows rushed on board with their helpless orphans in their arms,' reported the *Aberdeen Herald*, 'while parents and friends followed in equal grief.'

They frantically searched the ship for the sight of familiar faces, but all that they found were empty hammocks.

There was one last chapter in the story of the missing ships. None of the five ships that returned to port had had any contact with the *Swan* of Hull. By the end of March the vessel was still stuck in the ice—and only one cask of bread was left. Fourteen men volunteered to try to reach the Danish settlements at Four Island Point or Lively. On 1 April two men were seen about two miles from the *Swan*, standing on a hummock of ice. They were the only survivors of the party.

On 14 May two rescue ships were sighted. The *Swan* was towed into Whalefish Island Harbour and later sailed south to Hull, reaching the port on 2 July—the day that a memorial service was being held for the 'lost' whaler. Thousands of people turned out to greet her, cheering hysterically as she was towed up the Humber.

To-day, more than a century and a half after the *Dee's* escape from the 'breaking-up yard', there are still unanswered questions about the whaler's tragic homecoming. Why, for instance, was the *Dee's* distress signal ignored by the brig that they saw after escaping the ice? And what was the explanation of the behaviour of the crew of the fishing boat? One explanation given was that it was fear of the plague. It is said that the men who manned the fishing boat and refused them food and help were from Hoy. Flett believed this, and from then on had little good to say about anyone from Hoy.

Perhaps the most intriguing question of all was how Adam Flett survived that terrible voyage, when so many others died. To find the answer I went to the tiny farm of Nistaben, north of the Kirk on the Hill. The Harray man never returned to the whaling after his ordeal on the *Dee*, although he served on board the London mission ship *Harmony* when it carried missionaries and supplies to the Arctic. He sailed from Shields for a time, but finally came back to Orkney and bought the farm of Handest, near Dounby. Later, the family moved into the neighbouring farm of Nistaben, where his wife, Anne, had been born.

The two small farmhouses lie within a stone's throw of each other, looking as if they had changed little since Adam Flett settled there after giving up the sea. Handest had been empty for about six years, and it was at Nistaben that I met the whalerman's great-grandsons—three bachelor brothers, Bertie, Billy and Harry Flett. Bertie showed me a Certificate for Ability in Seamanship which Adam was given when he sailed to Labrador on the *Harmony* in 1851. For seamanship the marking was 'G' for good, while 'Character for Conduct' rated a 'VG'—very good.

The Flett brothers often used to sit at Handset and listen to their uncles speaking about Adam and the *Dee* disaster, telling stories that had been handed down through generations of the family. Now, *they* were telling them, and as I sat in that cramped, untidy room at Nistaben it was almost as if the old whalerman himself was talking. I heard about how the crew of the *Dee* had cut the ice around the ship to stop it from being crushed, about the men using wreckage from other vessels to brace up their own ship as the ice 'nipped' and squeezed it, and about how Adam had thrown his blankets on to the ice to get rid of the lice, only to find when he got into bed that they were still there, as pernicious as ever. I heard, too, about how he had brought back the body of James Tulloch on the *Dee*, with the coffin tarred inside and out, not only to preserve the body but to keep the smell down. He had promised that he would look after Tulloch's wife and family—and that is what he did.

Adam, said Bertie, was small and tough. He had an even temperament, he rarely worried—and he believed in keeping fit. While other members of the crew lay in their frozen bunks and gave way to despair, Adam was out on the ice running to keep warm. He told his sons about the moment when, as they drifted south locked in the ice, he suddenly knew they had beaten the 'breaking-up yard'. It came when he looked up and saw a lamp begin to swing—a clear and unmistakable sign that they were free of the ice and out into the open sea.

The years slipped away when Bertie Flett handed me a small, well-thumbed book, an original copy of *A Narrative of the Suffering of the Crew of the Dee, while beset in the ice at Davis' Straits, during the winter of 1836.* It was published in Aberdeen 'for the benefit of David Gibb, seaman on board of the Dee'. The date on it was May, 1837, the month in which the *Dee* arrived back in Aberdeen, but what caught my eye was the sloping writing on top of the page—'Adam Flett Handest Harray one of the surviving sea men on Bord the Dee'. It was Adam's own copy. He had left a message on it for any future readers—'Tak care of this Book.!' I closed it carefully and handed it back to his great-grandson.

CHAPTER SEVEN

Caa'in Whales

When I was at Olnafirth on the north-west coast of Shetland, Josie Manson, a one-time whaler, showed me a whale's ear-drum that he had in his house. It was like a huge sea shell, the kind you put to your ear to hear the sound of the waves. I remember old John Gray, in Stromness, telling me of the ear-drum that he had seen cut from a dead whale in the Antarctic. Whales whistle to each other, he told me. If you listened to the shell you could hear whale noises—'you get songs out of them, you see.' So I lifted Josie's drum to my ear and waited expectantly, listening . . ..

It may have been the wind groaning down the firth, or the lapping of waves along the voe, or just my imagination, but I was sure I could hear *something* coming out of the ear-drum. The whole mystery of whale sounds was locked in that shell. Sitting with it glued to my ear, I was, I suppose, hoping for some ghostly whisper from the past, for the ear-drum had been found on the shores of Olnafirth, not in the Antarctic, and it was from this long voe that Norwegian catchers had gone out to hunt the whale. Here, too, on the western hem of the Shetlands, the Caa'ing whales had been driven ashore and slaughtered.

Caa'ing or Pilot whales were not much bigger than the boats that chased them, and the Arctic whale hunters thought them too small to bother about. They must have seemed like sprats alongside the mighty Bowhead. The Greenlanders called them 'black dolphins', and the Newfoundlanders nicknamed them Potheads because of the oil-filled bulges on their foreheads. The hunting of Pilot whales can be traced back to the time of the Normans and Vikings, and perhaps even before that. Their killing was brutal and bloody. They were driven—or caa'd—into the shallows by shouting fishermen who banged their paddles on their boats and threw stones at the panic-stricken 'fish'. When the whales ran aground they were cut and stabbed to death with spears, lances or any other weapons the hunters could lay their hands on.

The killing of Pilot whales was described in lurid detail by visitors to the Shetlands last century, when this type of whale hunting was common. One of the chroniclers was Samuel Hibbert, a Manchester geologist who liked to tramp about the countryside in a battered hat and an old shooting jacket. He saw a Caa'ing whale hunt in 1818 when he was walking through Shetland with his dog Silly, collecting specimens, and he wrote about it . . . the 'work of death' as the fishermen struck at the whales with harpoons made from long iron-pointed spits . . . the 'dreadful convulsions' as the animals lashed the water with their tails . . . the sight of carcases strewn bloodily along the shore. 'The sun set upon a bay that seemed one sheet of blood,' he said.

The Caa'ing whales still haunt the shores of Orkney and Shetland, although the killing has long since ceased. They can be seen, especially in winter, off islands where they were once trapped and killed. Nowadays, as if in a belated act of contrition, the islanders tow off stranded whales, but it is often a hopeless task, for the whales have a herd instinct which makes them stick to their 'pilot'. When they were being massacred in the Shetland bays, the same sort of instinct made an escaping whale turn back at the last minute and meet its death with the rest of the pod. The Faroese called it the 'return to the blood'.

Although the killing of Pilot whales came to an end in the Shetlands in the early years of this century, Pilot whales are still driven ashore and slaughtered in the Faroe Isles. It is, say the islanders, part of their tradition and culture, and they claim that it is necessary because whale meat is essential to the Faroese diet—'Pilot whaling in the Faroe Islands has kept its original function as a communal, non-commercial whale drive for free food,' I was told. In this day and age, it is a sterile argument.

Shetlanders and Orkneymen who hunted the Pilot whale did it for oil, not food. They have never gone in for eating whale meat, except in times of dire need. Captain John Gray's grandfather ate it up in Papa Westray, as did other crofter-fishermen. 'They did it in those days because they were glad of it,' said John. Like his grandfather, John himself has eaten whale meat, and for the same reason—there was little else to eat. That was when he was on a supply ship in South Georgia. It was either whale meat or fish. 'We could have filled the boat with fish,' he said. Whale flesh was like horse flesh. 'Our boys couldn't cook it right. It had to be hung until it was nearly black—all the oil would drip out of it then. It looked grand in the pan, but we didn't like it very well.' The Norwegians, on the other hand, enjoyed it.

The whale-eating Faroese still occasionally take part in a traditional 'dance of the whales' after a drive. This has its origins in a ritual in which, wet and cold from the slaughter in the sea, they 'danced themselves warm', chanting traditional Faroese ballads. Nowadays, from what I heard, most people go home in their cars to change out of their wet clothes and wait for their share of whale meat to be allocated, but whether or not the dance had anything to do with keeping warm in the first place is doubtful. Long before the fishermen had cars they went home, changed into their national dress, and came back to the harbour-head to 'sing verse after verse of monotonous and seemingly tuneless ballads until their voices were hoarse.' It seemed to have more to do with primitive blood-lust than with keeping themselves warm.

Even to-day, children often help their parents to collect meat and blubber from the dead whales. They also like to collect whale teeth and 'enjoy examining the various parts of the whale'. This is regarded as part of their education, giving them an 'understanding of the source of their food'. It is certainly a far cry from the days when youngsters of ten years and upwards actually took part in the killing, slashing at the whales with knives lashed to sticks. Yet youngsters of the 1990's must still be deeply affected by such carnage, even if they don't take part in it. I couldn't help thinking of Samuel Hibbert's comment about the children of a century ago, who, with 'bloodthirsty exultation', added 'new tortures' to the whales' gaping wounds.

The Faroese call Caa'ing whales *grind* and their slaughter is known as the *grindadrap.* In 1986, partly in response to a sustained campaign by animal welfare groups, the government introduced new regulations to tighten control of the beaching and killing stages of the drive. Despite this, there has been no let-up in the flood of complaints. In 1991, the International Whaling Commission considered whether or not they should sit in judgement on small as well as large whales, but no decision was taken. It is still very much a live and controversial issue.

The killing of 'black dolphins' has taken place in most bays in Orkney and Shetland at one time or another. Women as well as men took part in the caa. There is a fanciful tale of how during one drive a woman in one of the leading boats, seeing the whales turn around and begin to make for the open sea, jumped on the back of the 'pilot' and guided it back towards the shore. They say it happened in Shapinsay, but how true it is is anybody's guess.

The whale hunt was an important event, coming before work, play, food, drink, the kirk—and even death. John Gray told me a tale that has become part of Orkney whale mythology. This was about a Westray farmer who was making a coffin for his newly-dead wife when he heard the call, 'Whales in the bay!' He immediately abandoned the coffin and his dead wife and made for the shore, explaining to his surprised friends, 'I couldna afford to lose baith wife and whales on the same day'.

The earliest record of a whale hunt in Orkney and Shetland goes back to 1691, when 114 were driven ashore near Kairston on the mainland. In Shetland, a record 'kill' of 1,540 was reported from Quendale in 1845. It is an astonishing figure, but it has to be remembered that as late as the 1840's shoals of a thousand or more were seen around Shetland each summer.

In 1831, over 300 whales met their end at Maill's Ayre, Cunningsburgh, and some six miles south at Channerwick

800 were killed in the following year. Less than a mile east of Channerwick is Hoswick, which won a small niche in both legal and whaling history at the end of last century. Houses huddle against the wind on the edge of the shore at Hoswick and a narrow road climbs up towards a headland with the curious name of the Berg. At the top of the brae you look down to where the bay meets the land in a long, wide 'U'—a perfect trap for caa'ing whales.

It was here, in September, 1888, that over 300 whales were driven ashore, killed, and auctioned for £450, which was no small sum in those days. Two landowners in the district, following their normal practice, claimed their share, but this time the fishermen refused. They were taken to court in Lerwick, where the Sheriff ruled that the landowners had no rights to a share in the profits. When the lairds carried their case to the Court of Session in Edinburgh, a fund was established and contributions poured in, one major sum coming from as far away as New Zealand. The higher court's decision was that all proceeds were to go to the crofter-fishermen—and none to the lairds.

Hoswick is one of a number of tiny communities scattered about the windy coastal strip between Channerwick and Sandwick. They have names like Stove, the Kirn, Golgo, Cumliewick and the Table of Stoo, Bannock Hole, Corbie Geo Cave, the Gun and the Stack of Billyageo. These are all within a few miles of Hoswick. The name that intrigued me most was Broonie's Taing, which I could see across the bay from the brae above Hoswick.

Broonie's Taing is at Cumliewick. I never discovered who Broonie was, but the 'taing', or point, was where the Norwegians once had a whaling station. There is nothing there now to show that it ever existed, but derelict warehouses and rusting oil tanks are ugly reminders of another oil operation that failed: North Sea oil, not whale oil. Hudson Offshore Enterprises built a huge complex at Broonie's Taing, and, instead of whale catchers, oil supply ships and dredgers lay within the extended breakwater. This offshore oil base closed more than ten years ago and some of the buildings were used by other firms after the Hudson Bay company pulled out. Now the deserted jetties and the empty warehouses have been left to the wind and the rain.

Norway became the undisputed leader of the whaling world after Svend Foyn's invention of the explosive harpoon in 1868, but overhunting brought a rapid depletion of whale stocks and the Norwegian whalers had to look for new fisheries. Whaling on the Norwegian coast was banned after fishermen had rioted and completely wrecked one of the whaling stations in the summer of 1904. The Norsemen, shut out of their own whaling grounds, turned their eyes towards Shetland.

The trouble in Norway was set to repeat itself in Shetland when Norwegian companies were given permission from the British government to establish whaling stations there. There were immediate protests from herring fishermen, who saw whaling as a death blow to their industry, and crofters also complained about the stench and pollution of the beaches. There were large anti-whaling demonstrations in 1907, and in the following year an appeal was made directly to Prime Minister Asquith.

Some Shetlanders, however, took the view that whaling gave employment at a time when many people were on the dole. Norwegian technique, capital and manpower had brought about a world-wide expansion of whaling, and it is unlikely that they were greatly perturbed by the Shetland protests. The Norwegians were never over-sensitive to public opinion; they had a Norse hardness about them. The pious Svend Foyn had invented the explosive harpoon (he became a wealthy man and left four million kroner to his favourite missionaries) and in earlier days the Norwegians killed their whales by infecting them with pathogenic bacteria and letting disease do the rest.

One of the first Norwegian whaling stations in Shetland was at Olnafirth, where Josie Manson has his fish farm. The Norwegians who went whaling from Olna must have felt at home in this wild landscape, sheltered from the Atlantic by the island of Muckle Roe. The Olna station was opened in 1904, along with Colafirth. Most of the stations had three or four catcher boats, each with a crew of about a dozen men. The manager at Olnafirth lived in a house on the site of Josie's present home. Some of the station buildings were still standing when Josie came to Olna, and behind his house are two links with its whaling past—the head of an explosive harpoon which still has the powder in it, and a large round stone like a millstone, which was used for sharpening flensing knives.

Away to the north the long elbow of Ronas Voe stretches west to a headland with the curious name of The Faither. It can be reached from the main road at Swinster or from Urafirth, which is not much more than a mile from Hillswick, where 360 whales were driven ashore in 1741, a year of famine. There is a pier at the end of the road and not much else, but it was here that the Norwegians opened the first of their Shetland whaling stations, two of them, in 1904. It is difficult to think of this peaceful inlet as a setting for tragedy, but in 1867 a steam whaler from Hull, with the figurehead of Diana the huntress on her prow, limped into Ronas Voe and brought to an end a horrifying episode in the history of whaling.

The *Diana* sailed from Lerwick on 8 March, 1866, bound for the sealing grounds off Jan Mayen. More than half the crew of fifty were Shetlanders, taken on as oarsmen and line-coilers for the whaleboats. The officers and harpooners were all Hull men, and the master was Captain John Gravill, a veteran of fifty years in the whaling trade, who had earlier commanded the Dundee whaler *Polynia*. They had bad weather and bad luck at Jan Mayen and in early April Captain Gravill abandoned the sealing and returned to Lerwick before sailing to the Davis Straits for the whaling season.

The *Diana* forced her way through Melville Bay, a notorious graveyard for whalers (in 1930, 21 ships were abandoned in the bay) and pushed on to the West Water. In Ponds Bay, she met up with ships from the Scottish fleet, all with gloomy reports of ice jamming the coast. At the beginning of August, Captain Gravill decided that any attempt at whaling was hopeless and turned for home, moving south in company with the *Intrepid* of Peterhead. Both vessels ran into trouble, but on 1 September the *Intrepid*, with her 60 h.p. engine, was able to force her way through loose floes, leaving the *Diana*, with her 30 h.p. engine, stuck in the ice.

With food running short, Captain Gravill took a gamble. He drove his ship into the pack ice so that it would carry the ship south into clear water. They crossed the Arctic Circle as winter temperatures began to bite. The surgeon, Charles Edward Smith, lined his bunk with canvas, sealskins and a piece of polar bear skin, but still the cold cut into him like a knife. He watched his breath form into tiny icicles on the nails around his bunk. His birthday came on 24 October, and he thought he would never see another one. Meantime, the food grew scarcer.

December came. There was plum pudding for Christmas, but little else. Scurvy began to take its toll, and even the surgeon found his teeth growing loose. Captain Gravill, old and ill, gave way under the strain and was found dead in his cabin. His body was sewn up in canvas and placed under a tarpaulin, and the *Diana* went on its way with a dead captain on the bridge.

The mate, George Clark, took charge, but it was the surgeon, Charles Smith, who kept the men going. When a Shetlander, blood running from his swollen gums, pleaded with him to be allowed to die in peace, Smith turned on him fiercely. 'You want to die, do you? No, by Jove, you shan't die!,' and he walked him up and down the deck for hours, persuading him that there was still hope, that the *Diana* and its crew *would* make it to Shetland. When they reached Ronas Voe, the man who had given up hope was one of the survivors.

January came, Medicine froze in the bottles and wine turned into solid lumps of ice. The surgeons dog, Gyp, had a fit and had to be shot. The last casks of beef and flour were opened. On 13 February the first man died, worn out by exhaustion and lack of food. He was Forbes (Purvis) Smith, a Shetland man, who had signed on at Hull exactly a year earlier.

March came—and another Shetlander died. But there were signs of spring and a linnet was heard singing in its cage. In mid-March the *Diana* reached clear water and the officers sat down to a celebration dinner of half-pound dumplings and raspberry vinegar. Ten days later, as the whaler ran south-east by south on the way home, the surgeon took stock of the crew's health:

> Thirteen completely disabled and confined to bed, 2 completely disabled with ulceration of feet, 14 severely affected, 6 getting very bad, 3 slightly affected, 5 very slightly affected, 3 show no symptoms of scurvy; total 46.

The toll, however, was still incomplete. Four more men died before they reached land—and three more died on the day the stricken whaler sailed into Ronas Voe. The *Diana* lay in Ronas Voe for a week and another three men died before a relief captain and a mate were sent out from Lerwick to take the vessel to port. Thirteen men, including the captain, lost their lives, ten of them Shetlanders.

Fifteen of the survivors belonged to Shetland, among them young Christopher Tait, a half deck boy. He made fourteen more voyages to the Arctic after the 1867 disaster, and he lived to be the last survivor of the *Diana's* crew, dying at Aith in Shetland on 29 February 1940, at the age of ninety-four. Another survivor, David Cobb, married a Shetland girl and died in his early seventies when he fell down a ship's companionway.

The *Diana*, the days of the caa'in whales, the Norwegian whale hunters . . . they are all fading memories now in these northern isles. Ronas Voe has been returned to its solitude. The only reminder of the *Diana's* terrible voyage is a small, simple memorial on the quayside at Lerwick. It was erected in 1890 by Alderman Frederick Smith, of Westham, brother of the whaler's surgeon. It carries the names of the captain, John Gravill, and the surgeon, Charles Edward Smith, and the inscription on it reads: 'In memory of the providential return of the whaler "Diana" of Hull 1866–67.'

CHAPTER EIGHT

'Too Much Cough'

The island of Kinatuk is a tiny pin-prick on the map of the Arctic. It lies in the Davis Straits, off the west coast of Greenland, and it was near this barren spot in 1856 that the Hull whaler *Truelove* made fast to an iceberg and sent a harpooner ashore to look for a young girl's grave. He found it in good condition, although it had been ravaged by nine Arctic winters, and he was able to read the name on the bleached, weather-beaten headboard—'Uckaluk'.

This lonely grave marked the end of a journey that began in 1847 when Uckaluk, a 15-year-old Innuit girl from Nyattlick in the Cumberland Straits, set sail for England with her 17-year-old husband, Memiadluk. They were taken there at a time when there was considerable concern over the pitiful conditions of the Esquimaux living under the British rule on the Baffin Bay coast . . . 'these poor Polar savages,' said the *Manchester Guardian*, were miserably treated compared to tribes living in east Greenland under Danish rule.

The whalemen called them Yacks. When Uckaluk and Memiadluk joined the *Truelove* they were filthy, covered with vermin, and coated with oil or grease. In Hull, they washed once or twice a day, dressed in new sealskin cloths, learned how to use knives and forks, and acquired a taste for cooked meat instead of raw beef. They were also put on show to the public. During their stay in England more than 12,000 people in Hull, Manchester, York and other cities flocked to see them—'Admission Sixpence, Schools and Children, Half Price'. In March, 1848, laden with gifts, they sailed for home.

It should have been a journey with a happy ending. Instead, Uckaluk died on the way home and was buried in her icy grave on Kinatuk. Memiadluk completed the journey to Cumberland Gulf and, according to William Barron, the young harpooner sent ashore to check on Uckaluk's grave, led 'a lazy life'. Barron, who later sailed on Dundee whalers, becoming a captain in his own right, said that Memiadluk was 'too idle to hunt or fish any more so long as the presents lasted which he had received in England'.

It was claimed that Uckaluk and Memiadluk were 'the only inhabitants ever brought to England from the Western Coast', which the whaling historian, Basil Lubbock, in his book *The Arctic Whalers*, took to mean the only Eskimos brought to Britain. This certainly wasn't the case, for nine years earlier an Eskimo called Eenoolooapik was brought back to Scotland by Captain William Penny on the *Neptune* of Aberdeen. In later years, a steady stream of Innuit were taken to this country by whaling captains.

Uckaluk, who was orphaned shortly before sailing for Hull, pleaded to be taken to England—'Uckaluk no father, no mother,' she said. But not every Eskimo wanted to go to Britain. Eenoolooapik, who was seriously ill with lung trouble during his stay in Aberdeen, was asked if he would ever return to Scotland. He shook his head and replied, 'Wyte you, wyte you, me takkou,' which was his way of saying wait and see.

'Too much cough', was Eenoo's verdict on Scotland, and he was not alone. Others suffered from 'too much cough', and there was an uneasy feeling among the Innuit that when their people sailed for Britain they might not come back. It was certainly true that a number succumbed to illnesses to which they had no immunity, and the British authorities discouraged such visits. Ock-o-kok, who was in Dundee from 1873–74. died before his ship took him back to Greenland. He was seen standing on the quay waiting for his whaler to leave because he couldn't stand the cold. John Sakeouse, who became something of a celebrity when he arrived in Edinburgh in 1816, was 'a good looking healthy young man,' but he caught typhoid and died in Leith in February, 1819.

Yet there were other Eskimos only too eager to see the far-away land of the whalemen. One of them was Aaniapik, whose adopted daughter, Coonook, an attractive girl with jet black hair, had captured Eenoolooapik's heart. Old Aaniapik, who was said to be tottering on the brink of the grave, was given the curious nickname of Commodore Timothy by the sailors. They dressed him in a blue jacket and canvas trousers, with a cocked hat decorated with red tassels and he strutted about like a proud peacock, playing on his friendship with Captain Penny. When he heard of Eenoo's adventures in Scotland, he made it clear that *he* would have no objection to going back to Aberdeen on the *Neptune*.

Eskimos from Baffin Bay were still arriving in Scotland at the end of the nineteenth century and in the early years of the present century. Among them were Olnik, who came in 1885 or 1886; Uno-Atwango, who seems to have been at both Dundee and Peterhead, and Shoodlue, brought to Dundee by Captain Milne in 1895. Shoodlue, or Sugar Loo, as some people called him, was a medicine man from Blacklead Island in Cumberland Sound, and there is little doubt that he cast a spell over the jute-producing, marmalade-making Dundonians. He was highly popular with them, laying on an exhibition of his canoeing skill in Dundee Harbour, and playing his melodeon in the Hillbank Hall. He also liked their marmalade—he won the title of Dundee's champion marmalade-eater.

It was a strange and often frightening world to many of the Innuit from Baffin Bay. When Eenoolooapik went ashore from the *Neptune* at the Castle of Mey in Caithness he saw a cow and pony grazing in a nearby field. He stooped down and stalked them, as if he was at home hunting. When he got close he went through the actions of firing an arrow; he thought they made a good target. Later, he asked what kind of deer they were, and if they were the same as Esquimaux dogs.

Shoodlue, the marmalade-eater, looked in awe and wonder at the grey tenement houses rising above the streets of Dundee and waited fearfully for them to come tumbling down on his head. He thought they were icebergs. Nouyabik, another Baffin Bay native brought to Peterhead in 1924, lived on herring, donned a suit (he preferred it to Eskimo clothes), and in May, 1925, was given his first ride in a motor car. He climbed into an old 18 h.p. Star and was driven off in a fever of excitement to attend a special meeting of the Aberdeen University Anatomical Society.

Eenoolooapik and John Sakeous made the greatest impression on the Scots. Eenoo was invited to a dinner party, given quite deliberately to see how he would behave in a civilised society, but he acquitted himself well. He had seen how his 'betters' conducted themselves in fashionable circles—and he was a good mimic. His manners were impeccable; he smiled and bowed and listened politely to their chatter, giving the impression that he was no stranger to such old-world courtesies. At first, he liked his food in a half-raw state, but later he refused it unless it was cooked properly—'oko too little', he said, or 'too little heat'.

The wary Scots had kept an eye on his behaviour, but in the end he taught his hosts good manners. At a reception on board a steamship, one of the passengers put his gold watch chain around Eenoo's neck and told the surprised Eskimo that he could keep it. Later, changing his mind, he tried to get it back and was chided by Eenoo, 'You give me to take from me—not good—Innuit no do that'.

Photographs taken at the time provided conflicting images of the Eskimos, or Esquimaux. The word Esquimaux is a corruption of Useuqemow, which means eater of raw flesh. The Eskimos called themselves Innuit, or 'the people' (Innuk when it was singular). A studio portrait of Ock-o-kok showed him surly and uncomfortable in a white man's suit, while Shoodlue was smartly dressed, handkerchief peeping out of his top pocket, his melodeon on his knee. A bearded Uno-Atwango was seen wearing sealskin Eskimo clothes and aiming his bow and arrow at some invisible foe. Eenoolooapik, a name which the tongue-tied Scots shortened to Eenoo, was presented as the perfect gentleman, hair neatly trimmed, all dolled up in a striped shirt, ornate waistcoat, and an eye-catching jacket with buttons and shawl lapels.

Meanwhile, the Eskimos back home were also adopting the style and habits of the white man. They dressed in European clothes and 'could drink rum and swear round oaths in English'. They were acquiring a veneer of civilisation, but their lives were still firmly anchored to the past. The men on the whalers returned to Scotland with contradictory reports about them. On the one hand, they were said to be docile, gentle people, friendly and hospitable; on the other, they killed birds by biting off their heads, ate blubber and even boot leather when hungry, and now and again committed barbarous acts of cannibalism. So what sort of man was the Eskimo, the real Innuk? Many of the whalers formed close relationships with the Eskimos, sharing their homes as well as their women, frequently taking Innuit wives. David Cardno, a whaleman from Peterhead, knew more about them than most. He lived with the Eskimos for many years, and recorded the lives of 'the people'.

The lure of the Arctic came early for David Cardno, whose father was manager of a whaling station in Cumberland Sound. In 1866, when he was only thirteen years old, he stowed away on the Peterhead whaling brig *Lord Saltoun*, bound for Cumberland Gulf. The Arctic bared its teeth at him on that first voyage. The crew of the *Lord Saltoun* had to take to the ice when a huge iceberg bore down on them as they lay trapped in the ice. It scraped along the vessel's port quarter, carrying away the stern davits and twelve feet of sail and bulwarks, but the whaler escaped serious damage.

The Peterhead ship had to winter at Niantelik when the ice closed in. Young Davie, playing cards with three shipmates on the messdeck, became involved in an argument in which strong language was used. His father overheard them, grabbed hold of Davie's ear, turned him over his knee, and gave him a sound spanking in front of the other men. The boy, humiliated, left the ship and moved in with an Eskimo called Murloo, who lived twelve miles away in an island igloo with his wife, Oashooka. He stayed with them for three weeks, sharing their food and joining Murloo on his seal-hunting expeditions—it was his first taste of the Eskimo culture that he was to learn so much about in later years.

In those early years, Cardno discovered that the Arctic took a cruel toll of the men who came in search of bone and blubber, the 'black gold' of the whale fish. In his first winter in Niantelik he helped to salvage provisions from the 300-ton barque *Dublin* of Peterhead, which was destroyed in a fire caused by a seal-oil lamp. He was there when three members of the *Lord Saltoun's* crew and two Eskimos set off on dog sledges for Kekerten Station, more than fifty miles away, to get badly-needed supplies. When they arrived at the station they discovered that one of their number, Peter Corduff, was missing.

The Eskimos retraced their steps and found him three miles along the route they had come, stumbling blindly through the snow and suffering from exposure. They took him to Kekerten, where they were alarmed to see that both his feet were badly frostbitten. The only thing they could do to save his life was to amputate them, although they had no anaesthetic. The 'surgeon' was John Bruce, a Peterhead cooper, and his instrument was a surgical saw made from a sealing knife. The operation was successful—Corduff lost his feet, but lived. Two other men, James Reid and James Kynoch, who took part in a second expedition, were less lucky. They were caught in a blizzard and buried in snow, their bodies lost till summer came. Their names were added to a simple cross on Norrie's Island, the last resting place of many north-east whalermen. The island was named after the first whalerman buried there—George Norrie, an uncle of David Cardno's father.

So, in his first fourteen months as a whaler, young Davie Cardno learned that life could be hard and uncompromising in the Arctic—and that death was never very far away. For over thirty years he sailed on Peterhead and Dundee whaling ships operating in the Greenland seas and the Davis Straits, among them the famous *Eclipse*, the *Polar Star*, the *Xanthus* and the *Jan Mayen*. It was on the *Jan Mayen* that Cardno almost lost his life when a whale he had harpooned overturned the whaleboat. The six men in the boat were thrown into the icy water. Three of them were able to reach the upturned boat when it resurfaced, and Davie was able to save a fourth man who could not swim. The sixth member of the crew disappeared, and neither he—nor the whale—were seen again.

In 1910, Cardno became manager of the two whaling and trading stations in Cumberland Sound owned by Robert Kinnes & Sons, of Dundee. He was at Kekerten for a year, and at Blacklead Island for another year, and he went back to Kekerten as manager in 1914. This time he had supplies for only twelve moths, but at the end of that period there was no sign of a relief ship, nor did it come in 1916. He lived on a diet of caribou, salmon and birds' eggs, and drank hot water instead of tea and coffee. For two years he was unaware that a terrible war was raging in Europe, and then the news reached him through a whaling station forty miles away.

Cardno passed the time playing solitaire and reading—and writing up his notes. He was the only European in a settlement of 300 Eskimos, and, however much he longed to return to his home in Scotland, he had been given a remarkable opportunity to study how the Eskimos lived. He saw them building their igloos, setting off for the hunt after a meal of raw flesh and water, tracking down a seal's nursing cavern and prodding a baby seal on to the ice to be killed by a blow to the head. While the men were away, the women went about their domestic duties . . . rolling up the bedding, tidying the sleeping bench, clearing hoar frost from the window, and sweeping the floor with a brush made from the wings of a duck or a raven. Tribal life was patriarchal. The top people were the Angakooeet, the chief conjurers, and after them came the hunters and the old man.

Cardno also dealt with the Eskimo's casual attitude to sex. 'The native attaches no more importance to the functions of sex than to those of eating and drinking,' he wrote. 'They are not without a sense of the fitness of things or some idea of personal modesty. It is the height of ill-breeding to stare, for instance, at anyone whilst dressing or undressing.'

On 17 August 1917, Cardno saw a two-masted schooner approaching Kekerten. He rowed out to the vessel, boarded her, and was given a piece of salt pork and a biscuit—'it tasted a treat', he said. His long vigil was over. To-day, the notes he wrote during his years at Kekerten are in the archives of Aberdeen University Library, along with the sketches and diagrams he drew. They were gifted to the University by David Cardno's grand-daughter, Mrs William Cruden, who lives in Cromwell Road, Aberdeen.

Mrs Cruden still remembers her grandfather coming home on leave to Peterhead, bringing gifts for the children, including Eskimo dolls. She says he was a very striking man—'a lovely tall, blonde man'. She still has three of his paintings. One of them shows the Peterhead whaler *Dublin* beached and blazing on the shores of Cumberland Sound. In the picture is the *Lord Saltoun.* A second painting shows the end of the *Jan Mayen*, which was lost in 1882 after its bows were badly damaged by the ice. David Cardno, who was on board the *Jan Mayen*, was transferred to the *Active* and landed in Iceland.

Descendants of the Innuit who were at Niantelik in Cardno's time, and of the Eskimos who came over to Scotland from Baffin Bay in earlier days, still live at Cumberland Sound, and they still talk of the whalers. In 1988 an historic park was opened at the site of one of the old whaling stations, and there were other plans in the pipeline. Dorothy Harley Ebner, a Canadian author who set out to collect reminiscences of the Innuit's links with whaling along the coast of Baffin Bay, wrote of the Cumberland Sound park and other schemes as 'timely developments'. There were Innuit, she said, who remembered that 'long before the white Canadians came, the Americans and Scots went through their ordeals up here'. There were still people alive who spoke of the time 'when the whalers and Innuit worked together, helping each other out'.

CHAPTER NINE

Gentlemen Whalers

Ecclesgreig Castle hides away on a wooded hill overlooking the village of St Cyrus, its decaying walls still carrying a hint of the elegance that once marked this south-east corner of the Mearns. Behind the shuttered windows, the turrets, and the neglected gardens are memories of days when the titled families of Angus and Kincardine gathered there for the shooting season. The castle and the 750-acre estate are situated a few miles east of Montrose.

In the private burial ground of the Forsyth-Grants of Ecclesgrieg there is a grave laid out alongside other family burial places in memory of Osbert Claire Forsyth-Grant, who, according to the inscription on the stone, was 'Lost in the wreck of the Seduisante in Hudsons Straits, 24th September, 1911'. But the Ecclesgreig grave was never used, for Osbert Clare's body was thousands of miles away—in a lonely grave on the shores of Nottingham Island in Hudson Bay, where he met his death.

From the garden of Ecclesgreig Castle, I watched the fishing boats fussing about off the coast. I was thinking of young Osbert Clare, son of Frederick Grant Forsyth-Grant, the Laird of Ecclesgreig. He had stood on the same spot nearly a century ago, looking out over the sea, thinking, perhaps, of the whaleships that sailed from nearby Montrose to the Arctic, or from Dundee, which was one of the greatest whaling ports in Britain. He lived on the doorstep of fishing communities like Johnshaven and Gourdon, and it was from old whalermen there that he first heard tales of the Greenland whale—and began to plan for the day when he himself would hunt it.

The family had a strong military background, but Osbert Clare was drawn to the sea. Although he got little support of encouragement from his family, he signed up for a voyage to Norway and Lapland in 1904, when he was twenty-four years old, and this taste of life in the Arctic reinforced his belief that this was where his future lay. In 1905, his father gave him money to buy a ship with which he could join the Dundee whaling fleet. He had little idea that the ultimate cost was to be his son's life.

His whaler was a tiny ketch called the *Snowdrop*. This Lilliputian vessel must have drawn disbelieving looks from Dundee's whalermen—it was the smallest vessel ever to sail with the Dundee fleet. The whales it set out to catch were almost as big as the *Snowdrop* itself. Hauling a whale on board was virtually impossible, and the first one caught had to be cut into pieces, an operation that took eight days instead of the usual eight hours. The *Snowdrop's* total catch on that first voyage was 'One Black Whale, two Walrus, 17 bears, 15 tons of oil and 17 tons of bone'.

Osbert Clare's first venture won him a headline in the local newspaper. 'Montrose Gentlemen's Trip—Perils of the Arctic'. For his second trip in 1906 the 'Montrose Gentleman' had as shipmaster a Captain Walter J. Jackson, from Dundee, who could never see eye-to-eye with the young owner. Osbert Clare, now known as Captain Grant, wanted to go north to the dreaded Melville Bay, the 'breaking-up yard', and then across Baffin Bay to Pond Inlet, but Jackson thought they were to be hunting walrus in safer waters. When the ship was taken to St John's, Newfoundland, for repairs after running aground and damaging here keel, there were arguments over the vessel's seaworthiness. The result was that Jackson and a number of the men packed their bags and went home. Later, two more of his crew deserted ship, so that Grant was left with three men to sail the *Snowdrop* back to Scotland.

Despite these setbacks, Osbert Clare returned to the Arctic in 1907. It was a bad year for Scottish whalers; there were no whales to be seen, and the *Snowdrop's* hold was filled with anything they could lay their harpoons and hands on—walrus, seals, foxes and Polar bears. It was said that half the country houses in Angus and Kincardine sported bear rugs from the Ecclesgreig 'gentleman's' haul.

In 1908, Forsyth-Grant recruited his crew from Aberdeen and Dundee, but one man, Alex Ritchie, came from the fishing village of Gourdon, near Osbert Clare's home at St Cyrus. It promised to be a profitable voyage. The *Snowdrop's* crew were helped by Eskimos from his shore base at Signia, Cape Haven, on the south side of Cumberland Gulf. 'Grant's people', they were called, and they joined the Scots whaleship *en masse* when it went fishing, sixty-five of them crowding on to the little ship. The men hunted and the women cleaned skins and hides. 'We filled our ship,' recalled Alex Ritchie later. 'We had 650 walrus and 600 seals and a great many bears.'

Then disaster struck. They were lying off Topjuack in Frobisher Bay when a storm blew up. The ship's log contained the terse entry; 'Ship dragging anchors. Strong S.E. gale. Struck the rocks with her keel. Got all the people ashore over the shore ice. Hurricane with blinding snow. Saved some sails and rigged tents'. The *Snowdrop*, battered unmercifully by the terrible storm, was a complete wreck. They were stranded in the Arctic with a dwindling food supply and only tents for shelter, plus igloos built by the Eskimos.

Some of the men set off in a boat to go to the head of the Frobisher Strait, and from there overland to Hudson Strait, where they might contact the Dundee whaler *Active* at Lake Harbour. They were forced to turn back, but ten days later they made another attempt. Again, they were unsuccessful. They then decided to walk about ninety miles to Cape Haven. There was practically no food left at Signia and they had to depend on what the Eskimos gave them, sometimes eating raw seal flesh. As time passed and the weather grew colder, they moved from their skin tents to snow houses.

In mid-December, when six Eskimo families were due to leave Signia to look for food and skins, Ritchie and five other men decided to go with them hoping to reach Hudson Bay. Forsyth-Grant asked Ritchie not to go, but he had made up his mind. In later years, when Ritchie was back in Gourdon and the *Snowdrop* incident had been all but forgotten, his son persuaded him to record the story of his Arctic adventure—a

500-mile journey across the ice on foot and by dog-sledge. The whaler's daughter-in-law, who to-day lives in the Borrowfield district of Montrose (her husband died eighteen years ago) still has a transcription of the tape. It is an incredible tale.

Alex spoke about his life with the Eskimos after leaving Signia, about their Hogmanay night on the ice ('A Happy New Year—I don't think!'), and about how his five shipmates eventually decided to turn back to the base at Cape Haven. One of them, Willie Morrison, was badly frostbitten and his toe turned black. When Ritchie asked the Eskimos what should be done about it they said they would take off the toe. Without chloroform it seemed a highly painful operation, but the Eskimos had a simpler answer—they would use a sandbag, give Morrison a bang on the head and make him unconscious, and then remove the toe. Not surprisingly, Morrison refused to let them touch him, but the result was that gangrene spread up his legs and after his return to Signia he died.

The Signia Eskimos returned to Cape Haven with Ritchie's companions, but the Gourdon man went on with another party from Hudson Bay. They camped at 'the foot of the place where the glacier breaks off into the sea' (this was the Grinnell Ice Cap, nearly 3,000 feet in height and covering 51 square miles) and then crossed the glacier. Ritchie's strength was giving out and the Eskimos left him—'I lay on the ice until I grew cold and then I would get up and walk until I would fall again'. The snow, blown by the wind, drifted like powder, and he thought, 'Now I am going to die or be frozen to death'.

Suddenly, as he lay on the ice praying, he heard voices. Two men, with a small sledge and seven dogs, appeared. They put him on the sledge and took him to their home, and after recovering his strength he again began the long, hard climb over the glacier. The Eskimos told him that if he could manage to stick it out he would be the only white man who had ever passed over the glacier. 'I never fell behind again,' said Alex.

When they came off the glacier, Ritchie stayed for a night in the Eskimos' village in the Saddleback Islands. Next day they set of for Lake Harbour, 100 miles away, and during the journey he fell ill, his flesh swelling and turning soft. 'I lay twenty-one sleeps that I did not know anything,' said Alex. 'They count their time by sleeps.' The *Active* was expected to arrive at the Saddleback Islands in the near future and Ritchie returned there with his Eskimo friends. One morning he awoke to 'the most beautiful music I had ever heard'. It was the sound of the rivers running. At the end of June, he saw the whaler *Active* lying at anchor off the shore.

Ritchie was so weak he could scarcely climb up the vessel's ladder. When he got on board he saw three men who had been with him on the *Snowdrop*. One of them, Jim Scott, ran for the ship's master, Captain John Murray, who looked at the dark, bedraggled figure in front of him and asked, 'Are you a white man?'. When he assured them he was white, Murray shook his head and said, 'What a sight!' He 'weighed seven stones, skin, clothes and boots'. Later, he sailed to Newfoundland on a ship called the *Lorna Doone*, and from there travelled home on a steamship, the *Siberian*, arriving in Glasgow in mid-November, 1909.

Eight days after Ritchie's arrival at the Saddleback Islands, a cable reached Scotland from Labrador with the news that Forsyth-Grant and the rest of the crew had been picked up at Cape Haven. They were taken on board a schooner which was bringing home the American big game hunter, Harry Payne Whitney, from a year's Musk Ox hunting in northern Greenland.

Alex Ritchie's daughter-in-law remembers him as a quiet man, who didn't smoke or drink. The only mementoes of his whaling days were a tusk behind the door and a large china dog given to him by Osbert Clare. He never returned to the whaling. The Laird's son, however, was still obsessed by the Arctic and in 1911 he bought a steam auxiliary, the *Seduisante*, whose name, appropriately, meant 'bewitcher' or 'seducer'. It was to be a fateful ship for the young whaler-trader.

Grant spend the winter of 1910–11 at Signia. His companion was an Innuit woman called Nangiaruk, the wife of Gotilliakjuk, who worked for the Scot. This 'borrowing' of a married woman was regarded as perfectly acceptable among the Eskimos. Grant was known to the Innuit as Mitsiga and it is said that Nangiaruk had children by him. Ainick, the son of Nangiaruk and Gorilliakjuk, is reported to have travelled to Scotland with him in 1907, but whether he ever saw Ecclesgreig is unclear. Forsyth-Grant, who was always a loner, frequently fell out with his crews, but he was always on good terms with his Eskimos, who liked and respected him. He seemed anxious to help them—'a gentle look came over his hard face as he discussed those he loved and admired'.

When the *Seduisante* arrived at Cape Haven, with a Peterhead shipmaster, Captain J. R. Connon, in command, Forsyth-Grant was ready for the hunt. He took his Eskimo 'people', some fifty of them, on board the little ship, as he had done on his earlier trip, and they set sail for an encampment at Topjuack. The *Seduisante's* course during its last voyage is uncertain, but it is thought that the ship was heading for Hudson Bay, where there were large numbers of walrus. The crew, at any rate, were unhappy at the thought of a lengthy journey in late summer, and there was even talk of mutiny.

The *Seduisante* was seen at Eric Cove, on the Ungava shore near the west end of Hudson Strait, and was told that there were walrus near Salisbury Island. 'That's for us,' Grant declared. On 14 September the Dundee whaler *Active* spotted the *Seduisante* some sixteen miles away, and a week later the *Chrissie Thomey* saw a schooner under full sail, driving through the Strait in blinding snow.

The captain of the *Chrissie Thomey* said they could see the vessel's lights, but they had no idea what ship it was. It gave them an 'eerie feeling', reminding them of the *Flying Dutchman*. 'Perhaps it was a phantom ship,' wrote the Captain. In fact, it was the *Seduisante*. What emerged later was that an Eskimo on the Scots ship had sighted a herd of walrus and Forsyth-Grant gave pursuit, ignoring warnings from the Eskimos that they were in a danger from rocks and shallow water. While steering a way through a narrow passage close to the shore of Nottingham Island, the *Sesuisante* crashed into a hidden reef. The Eskimos were ordered ashore and their lives were saved, but Osbert Clare and every member of the Scottish crew died. Only the bodies of the captain and the chief engineer were recovered; they were buried in a large pile of stones.

In years that followed, the story of the sinking of the *Seduisante* drifted into the realms of near-fantasy. There were highly coloured reports of mutiny on board, of quarrelling between officers and crew, of shots heard during the night, and of bodies washed ashore with bullet holes in them. The Eskimos on shore *did* hear shots, but they were thought to be signals asking for help. An attempt was made to launch a boat, but it was hopeless. The natives also said that when they were ordered ashore by Grant, the youngest crewman, Fred Livie, a 17-year-old deck boy from Dundee, tried to get a place in the boat. He was forcibly hauled back, and he wept.

No one knows what really happened as that stricken ship lay on a reef off Nottingham Island. Perhaps there was nothing more to tell. At any rate, it was said that Osbert Clare Forsyth-Grant became a legend in Baffin Island. To-day, even the legend has died. For many years a painting of the *Seduisante* hung in the Ship Hotel in Johnshaven, 'Granny' Hunter—Mrs Elizabeth Hunter—who was proprietor of the hotel for over sixty years, told me she got it from a local man, James Watt, who was a whaler, although he never sailed on the *Seduisante*. When she retired sixteen years ago at the age of eighty (she was ninety-five when I met her), she gave the painting to Lieutenant-Commander Michael O. F. Forsyth-Grant, Osbert Clare's uncle, the Laird of Ecclesgreig. The castle and estate have now been bought by a Swiss-American consortium, so the only reminder of 'Captain Grant' of the *Seduisante* may be an empty tomb in the castle garden.

Osbert Clare Forsyth-Grant loved the sea and adventure—and he loved sport. But the 'Montrose Gentleman' wasn't the only aristocrat with a double-barrelled name who joined the whale-hunters. Walter Livingstone-Learmonth, born in

Australia of Scottish parents and educated in Edinburgh, went to the Arctic ostensibly to map the coastline, but in reality he was looking for 'an unconventional life in quest of exploration and sport'. His only experience of life in the far north was a hunting and fishing expedition in Iceland in 1887.

When this globe-trotting Australian went to the Arctic he put himself under two of Scotland's top whaling masters. In 1888, he joined the Peterhead whaler *Eclipse* on a voyage to the Greenland Sea under Captain David Gray, and in 1889 he sailed on the *Maud* under the veteran Dundee master Captain William Adams. He had another reason for wanting to go to the northern waters; he planned to write a book about it. Captain Gray's son, Robert Gray, who was chief officer of the *Eclipse*, looked on the Australian with a slightly jaundiced eye. He said that Livingstone-Learmonth 'knows everything and has been everywhere. He knew how to shoot seals and bears before he had seen any of them'.

Livingstone-Learmonth had a curiously ambivalent attitude towards killing. He thought that a harpooned whale was 'a truly fine sight', and, writing about the 'grand sport' of Walrus-hunting, he declared that 'to see these huge brutes roll over is a reward worthy any discomfort to a healthy man'. On shooting Eider ducks, he said that the heavy thud of dead birds hitting the water was 'a cheerful sound'.

Yet he was horrified when he saw an Eskimo killing the 'Mollies' (Fulmar Petrels) that swarmed around the whale boats when a whale was being flensed. 'I saw one of the natives take up a position in the stern of one of the boats and catch the birds with his hand, killing them by biting their heads,' he wrote. 'The fellow had accumlated quite a heap of them before he was noticed, but this being a mortal sin according to a whaler's ideas, the man was soon routed out of the boat, and the poor birds were left to feed in peace.' He added that the Eskimo 'could have no object in destroying these birds except for the savage innate love of taking life'. This from a man whose 'bag' during the voyage of the *Maud* included more than 400 birds.

Nor was Livingstone-Learmonth too scrupulous about what came into his sights. He was on the *Maud*., he said, 'for shooting purposes', and he backed that up by shooting 26 walruses, 9 seals, and 4 Polar bears. There were also times when dead and wounded birds could be seen lying all over the ice or bobbing about in the wake of the ship. He shot at Polar bears and seals from long range, aiming at their chests and shoulders, with the result that wounded animals were left prowling about the ice.

Bear-hunting was common among whalers. Ship surgeons often took part in it, and the whalers themselves took home live bears to sell to zoos, or dead bears to sell to museums. In 1856 a bear could be sold to the zoological gardens in London for £35. In 1864, the Dundee whaler *Polynia* brought home 22 dead bears and 2 live. They were exhibited in the Greenland Yard at Dundee to raise money for the local hospital. Bear-hunting, however, was not simply a sport for captains and ships' surgeons. With the Bowhead whale brought to the point of extinction, bears became a prime commercial target for the whalers' guns. Nearly 500 bears were slaughtered by British whalers in Canada's eastern Arctic waters during the 1909–10 season.

CHAPTER TEN

Iron Maidens

The *Empress of India* lay in Lerwick harbour, the pride of the whaling fleet, bobbing jauntily under the eyes of the Shetlanders who watched her from the shore. This was the ship of the future, her bow twelve feet thick, carrying eleven boats, and built of iron to withstand the savage pack ice in the Arctic. Her Peterhead owners had spared no expense—there were graceful touches about the ship, brasswork in plenty, and a captain's gig that boasted a bronze bottom. Not only that, she had 110 men on board. Most of them looked down their noses at the old wooden whalers lying in the harbour.

The year was 1859, and it seemed as if Peterhead, the port that had dominated the whaling industry for nearly twenty years, was again leading the field. The officers and men on the *Empress* thought she would cut through the ice like a cleaver. 'All the crew expected they would make a small fortune,' wrote Captain William Barron, who was sailing that year on the brig *Anne*. 'They looked upon our sailing vessels with contempt.'

Sadly for the *Empress of India*—and for Peterhead's dreams of a new era in whaling—the Iron Maiden of the Greenland seas failed to live up to expectations. The first piece of heavy ice she came to stove in her port bow—and she sank within four hours. Her crew were save by the ships she had ignored.

It put an end to any idea that iron steamers could tackle the ice better than the old type of wooden whaler. That year, however, another new ship was launched that was to make history—the *Narwhal* of Dundee. She was the first auxiliary stream-powered whaler constructed of wood to be built in Dundee, and she set a pattern that was to make the city the premier whaling port in Britain. The firm that built her, Alexander Stephen & Sons, followed with the *Dundee* in the same year, the *Camperdown* in 1860 and the *Polynia* in 1861. By 1867, there were twelve vessels in the Dundee fleet.

It was an exciting period for Dundee, which needed whale oil for its expanding jute industry. But even with the coming of auxiliary steam whalers, the fear of being crushed in the ice never completely disappeared. Whaling was still a dangerous business. In 1884, four boats from the *Chieftain* were out hunting for Bottle-nose whales when fog descended, cutting them off from the parent ship. One boat became separated from the others and was picked up by a Norwegian schooner. The other three, one under Captain Gellatly, made a run of Iceland, 200 miles away.

The captain's boat was picked up by a Norwegian smack, and the specksioneer's boat reached land safely. The fourth boat had two men in it, Bain, the harpooner, and James McIntosh; the rest were boys. The boat drifted for fourteen days before being spotted by a shark fisher. The only survivor was McIntosh. The boys had died one by one, and then Bain had died. McIntosh, without food, thought that he might be tempted to cannibalise them and threw the bodies overboard. After being rescued he had to have both his legs amputated.

The ships built during Dundee's whaling years were known wherever whalermen met—the *Arctic*, the *Polynia*, the *Intrepid*, the *Aurora*, the *Terra Nova* and many more. The names of their captains became by-words in every house in the town, immortalised in old shanties sung ashore and at sea:

> There's the noble *Terra Nova*, a model with no doubt,
> The *Arctic* and *Aurora* they talk so much about,
> Art Jackman's flying mail-boat, the terror of the seas,
> Couldn't beat the auld *Balaena* on a passage fae Dundee.

There were men like 'Coffee Tam', Captain Tom Robertson, who got his nickname because he wouldn't allow alcohol on board his ship. Robertson, a Peterhead man, commanded the *Balaena* for a time and took her to Franz Josef Land to hunt Sea-horses—Walrus. In the late nineteenth century the killing of Sea-horses became popular because Walrus bull hides were greatly in demand for bicycle-making. Bicycle-makers paid 1/6 per pound for them. Their tusks fetched 2/6 per pound and the oil from them about £18 per ton.

While 'Coffee Tam' banned strong drink, Captain Charles Yule banned strong language. 'By the pipes of war!' he told his men, 'I am here to do my duty and I am going to see that you are helped to do yours!' His 'pipes of war' exclamation was the nearest he ever got to swearing. A strict disciplinarian, he was also very religious, hoisting the flag of Bethel on his ship when she sailed. He came late to the whaling business, having spent his early days trading in Australia and New Zealand, but after commanding a timber ship for Alex Stephen & Sons he took over the company's newly-built whaler *Esquimaux* and made his first trip to the Arctic in 1866.

The old hands regarded him as a novice, but in his first year he caught one whale when most of the Dundee fleet came home clean. The next year he showed them what the 'novice' could do, catching 7,000 seals in a six-weeks' trip and then going on to the David Straits to make one of the best catches of the season—nine whales. He took over the *Resolution* in 1880 and made his last voyage in the *Polynia* in 1883. He was known as Scotland's Grand Old Man of the Sea—and he lived to be 100.

One of the best-known and most successful masters in the last days of Arctic whaling was Captain William Adams, a 'kind, jovial, good-tempered man', who showed daring and enterprise. Whaling was in his blood and when he retired in 1884 he bought the *Maud* of Inverkeithing, converted her into a whaler, and sent her to sea under Captain John Watson. But that wasn't enough for the old whaling master; in 1887, unable to stay ashore any longer, he took command of the *Maud*. Three years later, after killing six whales, which produced 100 tons of oil and 117 hundredweight's of bone, Captain Adams was landed from the *Maud* off Thurso, seriously ill. He died before reaching Inverness.

The old century was dying, and there was little doubt that Arctic whaling was dying with it. Even sealing had become a profitless venture. The first decade of the new century accelerated the downward trend. In 1911, four Dundee vessels caught 7 whales, compared with 48 in 1881. Six whalers notched up a catch of 2,615 seals in 1911, compared with 152,706 thirty years earlier. In 1884, Dundee's most famous whaler, the *Terra Nova*, was launched, a symbol of hope, it was thought, but she was the last whaler built in Dundee.

Yet there were still men whose eyes turned to the North. In the year that the *Terra Nova* was launched, a 16-year-old Peterhead lad, John Murray, made his first trip from the Buchan port on the barque *Windward*. They were bound for the Greenland seas in search of the Right or Bowhead whale, but they caught only one Right whale, plus a few 'Botleys' and a number of seals. Young Murray was learning to be a whaler at a time when the Bowhead whale had become virtually extinct.

The Murrays were a whaling family. John and his older brother, Alexander, learned their skills from their father, Captain Alexander Murray, who took part in the Franklin searches and received a commemorative medal for it. All three Murrays served on the whaler *Perseverance*—and all commanded her at different times. When Alex and John joined the Hudson Bay Company they sailed together on the *Perseverance* to Repulse Bay, Alex as captain, John as second mate. The two Murrays spent the winter there, returning to Scotland in the autumn of 1893 with the blubber and bone of four Right whales in her hold.

They came together again when they worked for the Tay Whale Fishing Company, which was controlled by Robert Kinnes, who also set up the Cumberland Gulf Trading Company and the Robert Kinnes Company. Today, the latter still operates as Kinnes Shipping Ltd, serving North Sea oil operations. In 1907, John Murray sailed to Greenland as mate of the auxiliary barque *Scotia* under Captain Tom Robertson. The crew, despite 'Coffee Tam's' attitude to alcohol, drunk more than coffee on the trip, for some of them got hold of corn brandy and went wild. For a time they took over the ship.

The following year Murray was given command of the *Balaena*—the 'auld *Balaena*' of shanty fame:

And now that we are landed where the rum is very cheap,
We'll drink success to the captain for ploughing us o'er the deep
A health tae a' oor sweethearts and tae oor wives sae dear,
Not another ship could make the trip but the *Balaena* I declare.

The *Balaena* certainly provided John Murray with the success he wanted. In 1908 he followed 'Coffee Tam' to Greenland and came back with a good catch. From then until 1911, as master of the *Balaena*, he killed enough whales to keep the Dundee owners happy. He probed deep into northern waters, reaching 82° 20′ in Kennedy Channel. He could have gone farther, but he decided to turn back. 'I was looking for whales and not going to the North Pole,' he said.

John Murray became master of the ketch *Albert* in 1913 and took command of the *Active* after his brother's tragic death a year later. Alex Murray was looked upon as 'a good captain', and he was popular with the Innuit at Repulse Bay, whose lives revolved around the arrival of the *Active* each summer. The brothers were well-known all over the North. John Murray was nicknamed Nakungajuk—Cross Eyes—because he had a cast in his eye.

When they were wintering with the *Perseverance* at Whales Point in Roes Welcome, Alex had his wife with him. The ship was boarded by a crowd of Eskimo women eager to see their first white woman. They were fascinated by the balloon sleeves of her dress and the following evening they boarded the whaler again—all sporting balloon sleeves on their print dresses. Later, a child was born to Mrs Murray when they were frozen in at Chesterfield Inlet. She named the baby Percy.

Alex is said to have had a son and a daughter by an Eskimo woman. Dorothy Harley Eber, an American writer who set out to collect Innuit reminiscences of the whaling days, told in her book, *When the Whalers Were up North*, how she met an Eskimo called Leah Arnaujaq, who believed she was the daughter of Alexander Murray. She knew Cross-Eyes better, she said, but Alexander was her father. Eber also wrote about a hunter, Ikidluak, of Lake Harbour, who remembered the

Active arriving in Repulse Bay. He said his father was captain of the ship.

Alex never lived to see the war. He died on board the *Active* in the Ottowa Islands in November, 1912. Like the story of Osbert Clare Forsyth-Grant's death on the *Seduisante*, reports on Murray's death raised rumours and speculation. It was said that the cause was alcohol, which some whalers called 'forget-me-quick'. It was also said that the crew were afraid of him, and that they bound him up, presumably to stop him from going berserk. There was no mention of this in the log of the *Active*, which gave the cause of death as an 'internal tumour'.

There had been plans for John Murray to winter near his brother's ship with the *Albert*, but a sudden freeze-up forced him to stay at his anchorage in Welcome Sound. Later, however, he journeyed to the Ottowa Islands and took the *Active* home with its catch of six whales—one of the last great catches of the whaling years. His one and only trip on the *Active* before war broke out turned out to be an 'unfortunate voyage'. Not a single whale was seen and the trip was written off as a complete failure. 'The last of the Dundee whaling,' Murray called it.

John Murray's son, Austin, lives in Wormit, on the south side of the Tay, the river on which the whale hunters sailed out from Dundee. His father had a cottage on the shore, but Austin's house is high above the river. Standing there, looking out over the river thinking of the great whaleships that sailed down the Tay on their way to the Arctic, I remembered the entry I had seen in the *Active's* log. The date was 10 June 1903, the day before its departure for Hudson Bay:

> 2.30 Pm took in mooring. Pilot on board steamed down into Camperdown Dock, and out into the Tay. Nearly all the crew were drunk. Great number of people down giving us a hearty send off.

Next day, a gusting north-east breeze blew over the ship and the crew began to appear 'with swelled heads and on the hunt for a hair from the dog that bit them'.

Austin, who is now in his late seventies, went to sea like his father, becoming a captain on the Clan line. He never went on whaling expeditions, but when he was in the Royal Navy during the last war his ship was at the mouth of the St Lawrence River in thick fog. When the fog lifted, Austin could see forty icebergs. That was the nearest he ever got to the Arctic, but his father often told him about life in the far north, and about being frozen in.

He remembers hearing the story of an American whaler whose ship was trapped north of the Dundee vessel. The captain came across the ice to see Murray and asked him, 'How many men have you lost?'.

'I haven't lost any,' said Murray.

The American captain wasn't so lucky. He didn't have enough fit men in his crew to take his ship home, and he had no food.

John Murray always carried a dog sled on his ship so that he could hunt Caribou. He gave the American a few Caribou and a couple of bags of flour.

'He didn't understand why the Scotch bloke was doing so well,' said Austin.

He remembers, too, how his father came upon Eskimos who had never seen a white man. They were Sadlimiut, the original people of Southampton Island, who were called Pujait by the Innuit whalers. The word means 'dried-up oil', which was a derogatory comment on their appearance. When John Murray saw them they were up on a hill and scared to come down. He sent one of his Eskimos to speak to them and when they were welcomed and given gifts they agreed to come off the hill to meet the white men. One of them tried to rub the white off Murray's face to see what was underneath.

There was an old gramophone with a horn on board the ship, and the captain played a Harry Lauder laughing song. The Pujait were subdued and straightfaced when they heard the 'Ha! Ha! Ha's!' ringing out, but then they started to laugh themselves, rolling on the deck. John Murray left them to it, but later had a look at them through a porthole. He saw one of them peering down the horn to find out where the voice was coming from—or to see who was down there. Another Eskimo was taking a bite out of a record.

The Pujait had only bows and arrows with flint heads for their hunting. They were ravaged by influenza and died out in 1902.

During the First World War, John Murray was back on the ketch *Albert*, working for Henry Toke Munn, a wealthy trader and gold prospector, who had set up a syndicate to establish trading posts on Baffin's Island. He stayed with the *Albert* until she was sold in 1922, and on one of his voyages to Ushuadloo, north of Kekerten, the Eskimos provided a great feast and dance in his honour. The Eskimo women played the melodeon and an elderly Eskimo claimed he had Scottish nationality. The reason for this, he said, was that he had once made a trip to Peterhead.

For a number of years John Murray worked on Hudson Bay Company ships, the last being a small wooden vessel called the *Karise*. One entry in the log of the *Karise* was a signpost to the future—'Seaplane arrive to take out some of the furs'. It was a far cry from the days when the old sailing ships beat their way north to the Arctic, up by Godhaven and across the 'breaking-up yard' to Cumberland Sound. In September, 1932, Captain Murray took the *Karise* into Seattle and left her there. He was in his sixty-fifth year. He retired to Wormit and died in 1950 at the age of eighty-two. He was the last of the Scottish whaling masters.

Dundee hangs on to its whaling memories. There are still streets with names like Whale Lane, Baffin Street and Melville Street, and in a pub in the city centre, where old whaling pictures decorated the walls, I saw Captain William Adams keeping a stern eye on me from his Crow's Nest.

There are relics of the old whaling days hidden away in many Dundee homes. Austin Murray showed me beautiful Walrus tusks on which Eskimos had engraved kayaks, Reindeer and dog sleds, and he has an album packed with photographs taken by his father in the Arctic. The Broughty

Castle Museum has a flensing knife which has the initials 'W.A.' on it. 'W.A.' was William Adamson, who served on board a number of Dundee whalers and eventually became master of the *Advice*, which Captain William Penny commanded. The knife was presented to the museum in 1982.

The Broughty Castle Museum has one of the finest whaling displays in the country. There is a fearsome collection of weapons—hand harpoons, Svend Foyn's murderous explosive harpoon, boarding knives used to cut the blubber into manageable lengths as it was hoisted aboard the ship and big, stocky, brutal-looking whaling grenades, which were fired into the whale's body if the harpoon failed to kill.

A model of the *Terra Nova*, the last whaler to be built in Dundee, is on show. She was bought by the Government in 1903 to act as a relief ship for Scott's first Polar expedition, and Scott used her for his 1910 expedition. After that she went back to sealing in Newfoundland. She survived until 1943, when she was sunk by enemy action off the Newfoundland coast.

One of the exhibits in the museum has a curious link with the Franklin tragedy—a chunk of tobacco. It was found on Beechey Island in 1887 by Patrick Bell of the whaler *Resolute* of Dundee, part of a cache of supplies left by Sir John Franklin's expedition during his bid to find the North-West Passage. Another Arctic disaster is recalled by a flag with red, white and blue stripes. It was the Jack Flag of the ill-fated *Snowdrop*—the flag, carried by one of her whaleboats, which was raised to signal that it had harpooned a whale.

One of the photographs on display shows the football team of the Dundee whaler *Active* in 1894. The crews of the Dundee fleet were 'fitba' daft', and whenever two ships came together matches were arranged either in harbour or on the ice edge. In 1875, the crew of the *Victor* set up a game on the ice when their ship was held up in thick fog. The match was in full swing when an uninvited player wandered on to the 'pitch'. A full-grown bear had joined the game, chasing the ball as enthusiastically as the players. The men blew the whistle on the game, dashing frantically for the one ladder that hung over the side of the ship. The whaling master, Captain James Fairweather, showed the bear the 'red card'—he shot it.

CHAPTER ELEVEN

Blood Bath

A strange, weird world—silent, still as a dream, white and grey, painted with ghostly colours from a palette of pearl. No sound broke the long stillness but the sad hollow sigh of the great whales blowing their feathery jets of steam into the still, frosty air. Sometimes a muffled rumble came over the calm water as an ice-cliff broke and fell. The Snowy Petrels hunted in twos and threes along the ice edge, dipping into the dark sea for food, much as swallows do at home. The silvery yellow seals made no sound . . . .

That was how W. G. Burn-Murdoch, a scientist from University Hall, Edinburgh, described the Antarctic wastelands when he saw them from the deck of the *Balaena* in the winter of 1862. The 'auld *Balaena*' was one of a tiny fleet of four whalers that had gone south to find new fishing grounds in the little-known regions of the Antarctic. The other ships were the *Active*, under Captain 'Coffee Tam' Robertson, the *Diana* and the *Polar Star*.

The Greenland Bowhead whale population had been decimated, and now, with the Arctic seas swept clean, the whale hunters had turned their eyes to the south. The success or failure of their trip hung on a report by the explorer, Sir James Clark Ross, that in 1841 he had seen large numbers of black Right whales in the Antarctic, similar to the Greenland whales. The Dundee ships saw plenty of smaller Finner whales, but no Black whales. 'Finner whales of various kinds were seen every day,' said Captain Robertson, 'but we did not go south the catch "Finners"—they are plentiful nearer home.'

That was the end of the matter for the Scots, although Burn-Murdoch predicted that other ships would be sent out in the future and a station set up in the Falklands Islands. 'The seals alone would make such an industry profitable,' he said. 'Besides, there is still the prospect of meeting with the Right whale.' The Norwegians, on the other hand, had a more positive approach; if there were no Right whales there were Blues, Fins and Humpbacks. In 1904 they built a whaling station on the island of South Georgia and killed 183 whales in their first year of operation.

The British, who administered South Georgia and other island bases, set up a licensing system and put a tax on barrels of whale oil, but to evade this the Norwegians introduced the factory ship, fed by catcher boats. In the 1912–13 season there were sixty-two catchers at work in Antarctic waters, bringing off a 'kill' of nearly 11,000 whales. Between 1910 and 1940 the annual catch rose from 12,000 to 40,000 whales a year, with a peak of 55,000 in 1931.

Christian Salvesen & Sons, of Leith, whose whaling operations in Shetland had upset local fishermen, were looking to the Antarctic for new fishing grounds. Leith had been involved in Arctic whaling as far back as the early seventeenth century, and now Salvesen money and manpower were to be directed at what Burn-Murdoch had described as 'this strange Antarctic world'. In 1908, having been granted a Government licence, Salvesen sent an expedition to Leith Harbour in South Georgia—the start of an Antarctic connection that was to last for more than half a century.

They employed men from Shetland, Aberdeen, Dundee and Edinburgh. They earned their money the hard way, particularly those who wintered in South Georgia, maintaining and overhauling the catcher fleets. They lived in buildings made of timber and corrugated iron, each housing about fifty men in four-berth cabins. The sanitation in these foul-smelling barracks was said to be 'pre-historic'. Captain W. R. D. McLaughlin, an Aberdeen whaling master who was with Salvesen for twenty-three years, fifteen of them on factory ships, said that a shore-based whaling station was 'probably about the filthiest habitation of men the whole world over'.

The winters were appalling, bringing winds that were 'the fiercest and wildest in the world', with icy blasts reaching a force of 150 miles an hour. Corrugated iron from the roofs of buildings went sailing through the air 'like sheets of paper'. 'It was no wonder that all houses had to be anchored to the ground to prevent them being hurled into the sea,' wrote Captain McLaughlin. Snow lay on the ground in twelve-foot drifts, and avalanches crashed thunderously down from the towering peaks that overlooked the station.

In the early days, this dismal scene was made even worse by whale waste left lying on the sea-shore. The Norwegian station at Grytviken in South Georgia, set up by Captain C. A. Larsen in 1904, had miles of shoreline on Cumberland Bay choked with the bones of whales, spinal columns, loose vertebrae, ribs and jaws, and 'a hundred huge skulls within a stone's throw'. Robert Cushman Murphy, an American naturalist, who saw it in 1912, said that the district was 'an enormous sepulchre of whales'. No one, he said, could guess at how many thousands of flensed carcases had been carried out to sea.

When the scientist Dr F. D. Omnanney visited South Georgia in 1929, not much had changed. He saw beaches 'edged with ramparts of bleached bones, skulls, jaws, backbones and ribs'. Those were the days in which old John Gray, of Stromness, was making his first trips to South Georgia. He was second mate on the *Seringa* and he remembers the station at Leith being little more than 'a big concrete slip' on which they towed in the whales. If conditions on the shore were bad, they were not much better off-shore. there was no 'fridging' on the ships and the beef was kept in iced chests. 'Our food went rotten on us,' said John.

Nor did the passage of time greatly improve things. Neil McKichan, from Portincaple, Garelochhead, who made two trips to South Georgia between 1948 and 1953, says conditions there were 'very, very bad'. He has a clear recollection of the huts they lived in, and particularly of a 'new' brick building brought down from Norway. It was contaminated with bugs. Nor was the food anything to write home about, although the men were given a special treat at the week-end—'You got a whale steak every Saturday night for tea'.

Neil, who was a clipper, cutting up the dead whales, worked at both Leith Harbour and Stromness. He spent two winters in South Georgia, which put extra money in his pocket, but for that he had to endure the grim Antarctic weather, with little protection. He had to provide his own gear. Some men did two winters at Leith, one after the other, but the doctors on the station eventually put a stop to that because, as Neil put it, it 'affected their heads'. 'It's very, very lonely there,' he said. 'Very isolated.'

When Neil finally came home to get married, he had the job of looking after 200 extra passengers on the ship—penguins. They were being transported back to Scotland by Salvesen to be presented to Edinburgh Zoo. They were wild and strong and had big flippers that could break a man's leg—'no problem!' said Neil—and they had to be put into sacks on the voyage home. Salvesen shipped home over 800 penguins from South Georgia for the Edinburgh and London zoos, the first contingent in 1913.

Despite their leg breaking potential, the penguins were friendly enough on land, for Captain McLaughlin recalled in his book, *Call to the South*, how, while watching a football match between two teams of whalemen in South Georgia, he discovered that he had about a hundred penguin 'spectators' standing beside him. 'The penguins seemed as interested in the play as the humans,' he wrote. 'Every time the ball was kicked in their direction they moved back collectively and just out of range. As soon as play moved to another part of the field the penguins surged forward again. Their natural curiosity about human beings lured them frequently within arm's length—but just you try to catch one!' Captain McLaughlin believed that the penguins thrived in captivity. On one of his voyages he took home sixty penguins bound for one of the zoos.

The men who worked for Salvesen in South Georgia in the dying years of Antarctic whaling have been coming together again in recent years to talk over the old days. From his fish farm at Olnafirth in Shetland, Josie Manson, who did six seasons in that desolate corner of the world, organised a number of reunions of whaling men from South Georgia. There were 200 men, about 120 of them whalers, at the first reunion in 1986, and at a big get-together in 1991 they were presented with badges from the Edinburgh Salvesen ex-Whalers Club. Josie hopes to form an ex-whalers club in Shetland.

He comes from a whaling family. His father was bos'un on the factory ship on which Josie served, and his two brothers were on the same vessel. He was six years on one of the factory ships, driving the winch as a boy and graduated to the job of meat-cutter. Flensing was the same as in the old days, except that it was originally done off the side of the ship. Then, they took only the blubber; in the 1950's, when Josie was in South Georgia, they took everything except the guts. Flensing was the best-paid job. It was a simple job, said Josie, but it could be dirty—'the whole thing was dirty'. They used a very sharp knife, like a lance, and cut from the tail to the head. It was 'like peeling a big orange'.

There were about 1,500 men on the Leith shore station, half of them Norwegians, and half the British workers were from Shetland. The station used small catchers and ex-Naval corvettes—the corvettes were towing boats, used to tow whales to the factory ships so that the catchers could get on with the job of catching more whales. The year 1955 was Josie's last year at the whaling, and by that time there were signs that the days of the big whale hunts were over. 'The whales were getting smaller and smaller and more difficult to fish,' said Josie. 'Boats were getting bigger and faster.' Before he left South Georgia, they were fishing mostly for Finbacks—Finners—with some Sperm and Humpback whales. 'The biggest whale we got was a Blue whale,' said Josie. 'It was massive. When we got it up it blocked the whole ship for hours.'

This was unusual; the Blue whale, the largest of all the great whale species, had virtually disappeared from the Antarctic seas. Cold statistics show to what extent these giants were slaughtered. Before they were hunted out of existence, the number of Blue whales visiting the Antarctic annually totalled about 150,000. Between the two world wars, more than 20,000 Blues were killed each year, the figure rising to a record 29,400 in the 1930–31 season.

The Blue whale that Josie spoke about measured 96 feet. It must have been a mighty animal, a veritable Moby Dick, for only in exceptional cases did Blues reach a length of 100 feet or more. The fictional Moby Dick was based on Mocha Dick, an immense bull Sperm which attacked a whaleboat off the coast of Chile in 1810. E. J. Slijper, in his book, *Whales,* said that if Jonah had lived inside a whale's stomach it could only have been a Sperm whale, for the pharynx and the oesophagus of all other whales were too narrow to admit a man. Josie Manson argues that the Jonah story is fantasy. He showed me a photograph of himself sitting with another man inside a whale's mouth, perched on its lower jaw. You could, he said, just get your hand down the whale's throat.

Not all whalers were blind to the fate of the whale, or uncaring about the endless slaughter. Captain McLauchlin described his book as 'a plea by a whaling man who has spent many seasons in the Antarctic for action to save the whale before it is too late'. He said that the wholesale slaughter of whales was putting in peril the continued existence of the whaling industry.

During the 1960–61 season twenty-one expeditions operated in the Antarctic. The number of whales killed at sea was 38,812, and the production of baleen and sperm oil amounted to 2,146,472 barrels (six barrels to the ton). In addition, during the 1960–61 season Antarctic land stations at South Georgia killed 2,317 whales. When these figures were being written into the record books, Stewart Gray, from Burravoe on the Shetland island of Yell, was on a land station in South Georgia. Whaling stopped in the Antarctic in 1963, and he was there in its last year. He made two trips to the Antarctic, the first in 1946. There were few job opportunities in Shetland after the last war and it was normal practice for Shetlanders to go to sea.

But the main reason for signing on with Salvesen was simply the money; the whale oil boom was as lucrative to the men as the oil boom of to-day. 'A whaler was looked on as being very rich,' said Steward, 'especially if you did a season and stuck out the winter in South Georgia. If you did that, you were away for twenty-one or twenty-two months, and during that time there were no amenities, no social life. Conditions in South Georgia were very primitive. You didn't spend any money there.'

Two factory ships, the *Southern Harvester* and the *Southern Adventurer*, operated in South Georgian waters at that time, and each of these floating factories had twelve whale catchers. When the season finished on 7 April they tied up on each side of the pier at Leith Harbour and were serviced during the winter.

When Stewart did a winter in South Georgia, one of his companions was John Leask, a young lad from Scalloway in Shetland. He was lucky to be alive the following year, for in 1947, after wintering with Stewart in Leith, he became the only survivor of a Norwegian-built whale catcher, the *Simbra*, that sank off South Georgia. John, who was seventeen at the time, was deck boy.

Stewart, who had little good to say about the quality of the Norwegians' work, blamed the tragedy on the poor design and bad stability of the *Simbra*. When it happened, John was in the Crow's Nest spotting whales. If a whale was sighted, the look-out cried 'There she blows!' and held out his hand to indicate the direction it was taking. The catcher then went to 'full speed'. On this occasion, given a signal from the Crow's Nest, the helmsman put the vessel hard over. The *Simbra* was a slightly-built craft. Her stability at the time of the disaster was even more in doubt than usual because three days had passed since she had been bunkered at Leith. The result was that when the helmsman changed course the vessel went right over on her side.

The ratlines (the lines for climbing aloft) were horizontal on the water and John was able to make his way along them to a lifeboat launched by six Norwegians in the crew. He was only slightly wet, unlike the others. Three of the Norwegians died from exposure, one after the other. Three others died later. When the *Simbra* failed to make its routine call to the shore station, a search was organised. Just before darkness fell, John, the only member of the crew still alive, was spotted by a searching whaleboat and taken to the factory ship. When the whaling ended, the Scalloway lad emigrated to New Zealand.

Thirty years have slipped away since Stewart Gray came home from South Georgia for the last time. He still has vivid memories of his years in the Antarctic . . . of the isolation and the bad food, of winters when the cold bit into his very bones, and of the whale killing; more than anything, the whale killing. 'It was a blood bath,' he said. The gunners were poor marksmen. Some put five or six harpoons into the whale. 'When you kill a hot-blooded animal, what blood come out of it!' he exclaimed. He found it all very sad, almost beyond human comprehension that this kind of slaughter was allowed to take place.

At the start of the season in South Georgia, the whalers spent two weeks on Sperm 'fishing'. then the baleen fishing started—the Humpback whale, the Blue whale, the Fin and the Sei whales. They looked mostly for the larger whales, the Blue and the Fin. 'The biggest whale we caught in my time,' said Stewart, 'weighed 140 tons.' This was a Blue whale. 'Now they hunt the Minke whale,' said Stewart. 'We would never have looked at it.'

The 30 foot ten-ton Minke is the smallest member of the Rorqual family. In the 1950's it numbered more than a quarter of a million in southern seas and around 100,000 in the North Atlantic. It was hunted by the Norwegians and Japanese in the North Atlantic and at the 1986 annual meeting of the International Whaling Commission concern was expressed at 'the impending fate of the Minke whale'. When the larger species began to disappear from the Antarctic, whaling fleets began to hunt the Minke. From

1969 to about 1972, nearly 20,000 Minke were processed by Norwegian, Japanese and Soviet factory ships.

Postscript

The Future

What of the future? The Minke whale, which the Shetland whaleman Stewart Gray said they would 'never have looked at' thirty years ago, was in the sights of pro-whaling nations like Japan, the host nation, and Norway, for long an opponent of the 1985 moratorium, when the International Whaling Commission met in Kyoto, Japan, in May, 1993. They wanted the moratorium to end—and threatened to resume whaling without IWC consent if no agreement was reached.

The IWC estimated that there were 900,000 Minke whales in the world's oceans, with 760,000 in the Atlantic. The Japanese claimed that the western nations had made whales the 'sacred cows of the seas'—and that there was scientific evidence to justify resumption of whaling. In the end, however, the Commission rejected attempts to ease the moratorium.

Whatever the future holds for whales and whaling, man's influence on the wildlife of the Polar regions will still be felt, but in a vastly different way. The whale hunters have gone from both the Arctic and the Antarctic, but the whales, along with Polar bears, seals, walruses and penguins, face other threats. Pollution, fall-out from nuclear testing, raw sewage and chemical contamination . . . these are only a few of the dangers that have arisen in recent years.

One threat, ironically comes from people who want to 'shoot' the whale in a different way, with their cameras, for even the innocent presence of camera-carrying tourists hunting Moby Dick from the decks of tourist ships takes on a menacing note. The 1993 IWC meeting accepted a British proposal that a working group should be set up to assess whale-watching activities to boost tourism.

The British Antarctic Survey says that Antarctic tourism has caused disquiet among Antarctic operators. 'In some areas of the Antarctic Peninsula', the BAS reported in 1991, 'the number of tourists may be sufficient to have the potential to cause environmental impact at sensitive sites or in areas set aside for scientific monitoring.'

In the same year, Gilbert M. Grosvener, President of the National Geographic Society, sounded a warning about 'the tide of tourism now reaching remote places such as Antarctica'. With something like 3,000 visitors a year arriving in the Antarctic on cruise ships, he was concerned that uncontrolled waves of tourists could swamp 'this fragile unspoiled land'. In both the Arctic and Antarctic, whale-watching is becoming increasingly popular, and there are a number of travel agencies in Britain organising whale-watching holidays to the northern whaling regions.

People pollution in the Polar seas has not yet become a significant problem but pollution of a more deadly kind is said to be turning the Arctic into 'a deadly cocktail' of icy toxins and chemical pollutants. Early in 1992, Joseph Cummins, professor of genetics at the University of Western Ontario, Canada, made what he called 'a pessimistic prognosis'—that we might lose the majority of wildlife in the Arctic.

After his visit to Antarctica in 1991, Gilbert Grosvenor wrote in the *National Geographic Magazine* that painful lessons had been learned about the impact of thoughtless pollution by whalers, explorers and scientific stations. 'Wrecks of old whaling boats, heaps of ballast, rusted boilers and other trash testify to what we now consider a disregard for the environment,' he wrote. He said it was an attitude that could not be perpetuated by careless visitors to-day.

These unsightly reminders of the old whaling days can still be seen in South Georgia. Aberdeen, which sent its whaling captains to the Arctic last century, now retains a link with the Antarctic through the medical facilities it provides for scientists operating in the southern Polar regions. Dr Graham Page, head of the emergency department of Aberdeen Royal Infirmary, who served in South Georgia in 1988, showed me